OUR GREAT CANAL JOURNEYS

OUR GREAT CANAL JOURNEYS

A LIFETIME OF MEMORIES ON BRITAIN'S MOST BEAUTIFUL WATERWAYS

TIMOTHY WEST

JOHN BLAKE

Published by John Blake Publishing Ltd
3 Bramber Court, 2 Bramber Road
London W14 9PB, England

www.johnblakebooks.com

www.facebook.com/johnblakebooks ⬛
twitter.com/jblakebooks ⬛

This edition published in 2017

ISBN: 978 1 78606 511 7

British Library Cataloguing-in-Publication Data:

A catalogue record for this book is available from the British Library.

Design by www.envydesign.co.uk

Printed in Italy by Stige Arti Grafiche

1 3 5 7 9 10 8 6 4 2

Papers used by John Blake Publishing are natural, recyclable products made from
wood grown in sustainable forests. The manufacturing processes conform to the
environmental regulations of the country of origin.

Every attempt has been made to contact the relevant copyright-holders,
but some were unobtainable. We would be grateful if the appropriate people
could contact us.

John Blake Publishing is an imprint of Bonnier Publishing
www.bonnierpublishing.com

For Sam and Laura,
who love the canals

CONTENTS

© Getty

FOREWORD

BY PRUNELLA SCALES

I THINK THIS BOOK is a bit like one of our canal journeys; meandering along through our lives until it is suddenly carried away by a current, or a weir, or a sandbank.

I am married to a restless spirit. Tim doesn't like to stay too long in one place. This means that while theatre audiences in the provinces have been able to enjoy his performances in Chekhov, Sheridan, Ibsen, Brecht, Shaw and Arthur Miller, those in London, where it's supposed to count professionally, haven't.

Is this important? I don't know. What matters to me is that I've shared his passion for exploring unusual places, meeting new people and doing surprising things; that's been my life, and I wouldn't change it.

Most people seem to be able to enjoy a very nice, planned 'Annual Vacation'. Not us. We've just grabbed time when we could, gathering up the kids in their school holidays, and somehow we've managed the trade-off between leisure and work. Canal boating, however, initially couldn't have been further from our thoughts: this book tells how the idea suddenly became an integral part of our lives, and eventually led to the creation of *Great Canal Journeys.*

When we first started filming the programme, I wasn't quite sure what I was supposed to be doing. 'I'm an *actor*,' I told myself. 'I like playing people who are infinitely more intelligent than me, and say things considerably more interesting. I don't just want to be *me*.'

Tim doesn't seem to mind; he's done this sort of thing before. And actually, when I got into the rhythm, I really enjoyed it, and the waterways started to work their magic. On a canal, you can relax, enjoy the scenery and the wildlife, think about where exactly you are and let your mind wander back through the past.

I like to be given occasional reminders of my childhood. I was brought up in the country; I remember walking miles in the Lake District to bring down a baby lamb who had been born too far up the fell; I can still smell the paraffin from the oil lamps that we had to fill every night, and trim the wicks. We had lovely dogs. Viewing the pastoral scene from water level, at four miles an hour, gives me the opportunity to piece bits of memory together.

Before writing this foreword, I looked at some film excerpts we have of the canal journeys, and picked out bits that hold a special significance for me, because nowadays I don't remember things very well. The romance of Venice, the excitement of India, the beauty of the Midi, and all the much-loved domestic waterways: the Llangollen, the Kennet and Avon and of course the Oxford canal, where we first got together more than fifty years ago, when we were going round the country doing different plays.

In this book, Tim talks a little bit about our professional lives, but doesn't really dwell on it; it's not really within the scope of the book. I'm afraid that, since the onset of what we like to call 'my Condition', we no longer go to the theatre a lot, because I just can't remember much of it afterwards; so it's a bit of a waste. The cinema is much the same. Music is rather different; we can go to concerts, and even the opera, and I can come away spellbound, with neither of us needing to talk about what we've heard.

Social occasions are more tricky: at a party, someone may be telling me about his mother's death in a road accident, and I'll be properly sympathetic and then go away and talk to somebody else. A little while later I'll be back with the original person, and ask him for news of his mother. Patiently he'll tell me the story again, and I'll repeat my sympathy. But if there's a *third* time . . .

I'm afraid this is the reason we don't get asked out as much as we used to!

How do I feel about being in this situation, people ask? (Or, indeed, are hesitant to ask.) Well, angry, of course. I hate the idea that the world is going on all around me, but that so much of it is closed off. I soon forget my anger, though, as I forget nearly everything else.

I don't really want to talk about it. Instead, please enjoy this wonderful book, which relives our journeys together – both literal and metaphorical. I did.

INTRODUCTION

WHEN I WAS ABOUT FIFTEEN I went to stay with some friends in Bristol, and we saw in the local paper that a public meeting of the Kennet & Avon Canal Society (not a Trust in those days) was to be held beside a disused lock in the centre of Bath. We didn't know anything about it, but we had a free afternoon and thought it might be fun, so we went.

There was quite a crowd there, including the Bishop of Bath and Wells and the MP Chuter Ede. There were a number of young men holding placards saying 'SAVE THE K&A', and someone told us that this important canal, built by John Rennie and opened in 1810, had been gradually run down over the years until it was useless and derelict; and a lot of people thought it was time to do something about restoring and reopening it.

There was an opposition group, headed by a successful-looking farmer in a very nice tweed suit, who, in order to give himself extra height over the assembled company, had chosen to climb onto the upper gate of the abandoned lock and speak from there. Were the canal reopened, he complained, its path would lie across valuable land, which he could put to good use. He went on about this for rather a long time, and people were getting bored, so one of the young supporters had equipped himself with a barge pole, and with the end of it began to nudge the farmer's legs towards the far end of the gate.

Everyone had stopped listening to him by this time; instead we were fascinated by what seemed likely to happen next. There was no water in the abandoned lock, but it was full of mud. Surely, for his own sake and that of his lovely suit, it would be good to shut up and edge his way back onto firm land? But no. He went on talking, kicked out at the offending pole, lost his footing – and, yes, into the mud he went, and the *Bristol Evening Post* got the photograph.

That enjoyable afternoon prompted in me the first stirrings of interest in the subject of canals. I think it's always very easy to maintain you were born at the *wrong time* for things: too early, and you won't ever understand computers; too late, and you'll have missed out on black-and-white films. But, if canals happen to be your interest, well, then, you've been born at exactly the right time: there are still a few working-boat people around providing a direct link to the Industrial Revolution, while, at the same time, you can see a future opening up with more and more instances of canal restoration.

Each one of this country's canals will have its own history, determined by lots of things: the contour of the land, the cost of construction, the skill and imagination of its engineers and the level of demand for its freight transport. For centuries, canals were so much preferable to the cart tracks that served as roads; but then, later, the railways came along and provided fierce competition, frequently buying up the canal in order to let it perish through lack of maintenance. By the mid-1960s, commercial traffic on the waterways was virtually at an end, and boats were being sold as scrap.

Putting our canal network back into use is indeed a fairly recent idea. In 1946, the Inland Waterways Association (IWA) was formed by two men, Robert Aickman and Tom Rolt, to bring the situation to the notice of a wider public.

I never knew Aickman, but I did briefly say hello to Tom Rolt one wet afternoon beside the track of the narrow-gauge Talyllyn Railway in Merionethshire (he was campaigning

Robert Aickman (left) travels from Birmingham to the lower reaches of the Severn on the new floating headquarters of the Severn Wildfowl Trust.
© PA Archive

to save that, too, and famously succeeded). A professional engineer, but also a writer of distinction, he had acquired an old horse-drawn barge, installed an engine and converted it to living accommodation. He recorded his maiden voyage up the Oxford Canal from Banbury in the book *Narrow Boat*, which motivated a growing public determination to save what was still navigable of our canal system for use as a leisure resource.

Tom Rolt, apart from canals and light railways, took an interest in old road vehicles and racing cars, as well as pretty well everything emanating from the Industrial Revolution. He was keen to preserve things, but only if they *worked*. It was not enough for a thing to be beautiful, he said – it had also to be efficient. I admire that.

Tom died in 1974, but I was privileged to know Sonia, his widow, who during the war had joined the women volunteers who kept the canal boats running between London and the Midlands, carrying coal in one direction and essential engine parts in the other. Steering two heavily laden narrowboats, one with a motor towing the other – the 'butty' – was an arduous and often dangerous occupation, but she continued at it after the war was over, and her concern with chronicling past canal life, in photographs and written testimony, continued right up until her death a few years ago.

While the IWA's determination to preserve our national heritage met with success in some individual cases, it was up against a government policy to abandon totally half the existing waterways, drain the land and make it available for building. It was quite a long time before public concern,

1944: A full cargo of coal is transported by barge along the uncongested Regent's Canal during war-time, when all methods of transport were employed. © *Getty*

mainly through their MPs, was able to curb the destruction, and gradually to persuade the authorities that our canals were in fact an environmental asset. Eventually, the campaign led to three thousand miles of waterway being once again available to boaters, walkers and cyclists, with all sorts of fringe benefits to those living on or near a canal.

I, for one, am very thankful for it – and, reader, if you've ever been so fortunate as to find yourself with the time to glide gently along one of these beautiful waterways, drinking it all in, I suspect you are, too.

PART 1

MR &
MRS WEST

IN THE MIDST OF THINGS, two young actors met in a gloomy church hall in Pimlico, in order to rehearse a romantic television drama about the eighteenth-century Earl of Sandwich. It was a terrible play, but it was work, and in those early years television plays of any sort were thin on the ground.

As I was playing only a small part, I knew there'd be a lot of waiting around, so I had brought along the *Times* crossword. Across the room I saw that the attractive, rather soberly dressed young woman playing the Bishop of Lichfield's daughter had decided to do exactly the same thing. Her name was Prunella Scales. We began to sit together when we weren't needed, and jointly attack the crossword.

Two weeks passed like this, and then it was the recording day. We all turned up at the BBC Lime Grove Studios, put on our wigs and costumes and waited to be called to the set. Time passed. We were told that, unfortunately, we had hit the middle of an electricians' strike, and until this was settled there could of course be no possibility of recording the play. The negotiations went on until it was now clear that we'd run out of time, and the director called us together and told us that we could all go home. It had been found impractical to reschedule the programme, so we all made noises of mostly spurious disappointment and got out of costume. I turned to Pru and suggested we might go to the pictures.

So we got a bus to the Odeon Marble Arch and saw *The Grass is Greener*, starring Cary Grant and Deborah Kerr.

I don't think we were aware of the aptitude of the title. I was just coming to the end of an unsuccessful seven-year marriage, and my wife Jacqueline and I had finally agreed that we'd be better off apart, and free to explore other possible relationships.

However, when the film was over, all that happened was that Pru and I said goodbye, swapped addresses, and hoped very much we'd work together sometime. Pru went back to the Chelsea flat she shared with three other girls, and I returned to our family house in Wimbledon.

Some weeks later I got a postcard from her, saying that the BBC had resurrected that awful play and that she had reluctantly agreed to be in it again. Pru said she was sorry to find I wasn't there. I wrote back to explain that I was now engaged on something else. She said she missed me. I said I missed her too.

We began writing to each other fairly regularly. We've both always preferred the letter to the telephone, and, now that we were engaged in two different touring productions, we got into a rhythm of entertaining each other weekly with news about what the town was like, the theatre, the landlady.

Notepaper was expensive, and Pru took to using the backs of old radio scripts.

'They're very absorbent, aren't they? Do you think what you save on paper you lose on ink?' she wrote to me.

Chatty, jokey and inconsequential as these missives were, we were both accomplished enough letter-writers to be able to read between the lines. We were falling in love. So somehow we had to get to *see* each other; and finally, one week in the early summer, that became possible. Pru had gone straight into rehearsing a new play at the Oxford Playhouse, while I was on tour with the farce *Simple Spymen*, playing a week at the New Theatre round the corner.

Pru of course was rehearsing all day, and I was performing in the evening, so we didn't meet properly until Sunday morning. The day dawned warm and cloudless; we wandered round the city, had a nice lunch and hired a punt at Folly Bridge. I handled the pole dutifully for the first half-hour, ice-cold water running up my arms and soaking my shirt, then I sat down and just let us drift into the bank.

I still remember that afternoon on the river. I recall clearly what we were both wearing, the colour of the cushions in the punt, the parade of ducklings striving to keep up with their mother, and our being stared at critically by two moorhens in the reeds.

Bird-life is happily ever-present when boating and, to this day, watching my children feed the ducks brings back happy memories.

Pru and I have always loved the water – being on it, beside it, and sometimes *in* it, although I'm a terrible swimmer. Yes, we owe a lot of our lives to the water.

After finishing the play at Oxford, Pru joined the cast of *The Marriage Game*, which again meant a prior-to-London tour, while I was still going round the country with *Simple Spymen*. We were desperate to get together whenever we could, and we contrived a series of Sunday assignations at cross-points in our touring schedules. If Pru had been playing in Birmingham, and was about to move on to Newcastle, and I had just left Liverpool and was going on to Hull, then we might meet in Sheffield. If she was travelling south from Leeds to Brighton, and I was going in a westerly direction from Norwich to Bristol, we could perhaps rendezvous in Coventry. We had a lot of energy in those days; and we learned an enormous amount about trains.

Pru and I had grown up in very different surroundings, which meant that there were always interesting and surprising things to learn about each other's early life. Pru's father had been a professional soldier, who had fought in both world wars. Her mother had been on the stage but had given it up on getting married. The family lived in Abinger in Surrey, where Pru was born. She went to a girls' school in Eastbourne, which, when war broke out, was evacuated to the Lake District. After leaving school she enrolled at the Old Vic Theatre School in south London, where she was spotted by a rather good agent who put her up for a number of things. She auditioned successfully for a part in Thornton Wilder's *The Matchmaker*, opened in it at the Theatre Royal Haymarket and then went with it to Broadway. From then on, her professional life (we both hate the word 'career') embraced theatre, television, film and radio, and continued pretty well without a break.

My parents were both 'in the business', though my mother gave it up when her second child, my sister, was born. My father was a popular and respected actor, but it was a long time before he reached the West End, and most of my young life was spent chasing him round the country. He had just secured a job in Bristol, when war broke out, and he had to enlist as a War Reserve policeman. We got a flat, and lived there for the duration. In 1946, we moved to London (well, the London *area*) and, after attending the John Lyon School and the Polytechnic in Regent Street, I took jobs first as an office-furniture salesman and then, because I love music, as a recording engineer for EMI. Eventually, I owned up that what I really wanted to do was to be an actor.

I took the long and rocky climb through various different repertory companies of uneven quality. At the beginning I was in places that did *weekly* rep: performing one play in the evening while rehearsing the next one during the day, for perhaps forty-six weeks a year, including the panto. Obviously, we can't have done the plays very well, but we *did* them, and it delivered a form of training unavailable to young actors today. No drama school can possibly hope to provide that variety of text, style, manners, dialect, costume and behaviour that was needed if you wanted to keep your job in weekly rep.

So you could say I learned my trade by doing it, whereas Pru's drama-school training was rather more innovative and academic. When we first knew each other I thought of myself as a useful, quite versatile actor who could play older parts more cheaply than the more senior actors who ought really to have been playing them. Pru got me to think more imaginatively about myself, and about the real process of acting. That was perhaps the first of many occasions in life when she has seen fit to offer a subtle push in the right direction. I sometimes don't realise it till later.

Our lives have often followed wildly different directions, but we cherish the opportunities we've had to appear on stage together: *When We Are Married*, *The Merchant of Venice*, *Love's Labour's Lost*, *What the Butler Saw*, *The Birthday Party* and *Long Day's Journey into Night*.

Sometimes we feel, though, that our joint appearance could actually prejudice the play: if the characters we're playing are going through some romantic turmoil or insecurity, might the public say, 'Oh, no, it's all right *really* . . . they're happily married'?

My tour lasted a full twenty-six weeks, but, when it finally came to an end, I decided to audition for what was known as the BBC Drama Rep – and was accepted. The Rep was a company of about forty actors, paid a weekly salary to perform in all aspects of BBC Radio – plays, stories,

Pru and I celebrate an opening at the Grand Theatre, Blackpool, with inadvisable quantities of champagne.

letters, poetry programmes, comedy shows and the odd bit of announcing. I have always loved working in radio – for one thing you can play parts for which you're physically quite unsuited – and I passed a very happy, wonderfully varied year. I learned a great deal from my Rep colleagues, who included some of the great names of radio, and with them I enjoyed many riotous sessions in the BBC Club round the corner.

The significant thing was that I was being paid regularly, by the week, and so I had begun to embark on the serious process of getting a divorce. I had feared this would prove a wearisome and ignominious business; but, as it turned out, not at all. I was being the guilty party, and Jacqueline's solicitor had engaged an enquiry agent, who just wanted

My beautiful bride and I, newly wedded.

to know where and when he was supposed to discover me *in flagrante delicto* with 'Miss X'.

I was staying in a hotel in Cheltenham, where I was directing a play, so he would have to come up there. We suggested a suitable date, and Pru, as Miss X, came up from London to join me overnight. However, the agent didn't relish an early-morning start, and the train he proposed to catch from Paddington would not deliver him to the hotel until after I'd had to leave for the theatre and Pru had gone back to London. What could we do?

There was no problem, the helpful chap explained: Miss X's presence was not obligatory, nor, apparently, was mine. 'Twin indentations on the pillow will suffice,' he said, 'though perhaps an item of lady's night apparel might serve to clinch matters.'

I looked up a more acceptable train for him, we exchanged cordial wishes, and Pru went out to Marks & Spencer and bought a matter-clinching class of nightgown. The following morning I draped it over the bed, thumped the pillow twice, and went to rehearsal. In due course, the *decree nisi* came through.

One day, according to Pru, we were held up at some traffic lights on the A30, when I reached into my pocket and brought out the ring I'd bought in a rash moment in the Brighton Lanes, and asked, 'Will you marry me?' Obligingly, she said oh yes, all right, then the lights changed and we moved on.

We got married quietly at Chelsea Register Office on a Saturday morning in October. We both had to get back to work on Monday, but for the time being we had two nights of wedded bliss to spend at a riverside hotel in Marlow, which turned out to be possibly the most celebrated venue for illicit weekends in the Home Counties, with a waiter who gave me a conspiratorial wink and murmured, 'I think *the lady* has already retired upstairs, sir.'

The following spring we found we had a week off, and so we went off to Llangollen in North Wales for a week, and decided that was our honeymoon.

Our son Samuel was born in 1966 at the rather smart Queen Charlotte's Hospital, Hammersmith; by the time his brother Joseph came along, our gynaecologist had moved to King's College Hospital in the slightly more downmarket area of Denmark Hill. Pru remembers the marked contrast in the way the births of the two boys were registered:

Following Sam's birth, a very elegant man in striped trousers and a dark jacket knocked softly and came in. 'Good morning, Mrs, er . . .' – he looked at his list – 'West. Many congratulations on the birth of your, um, son. When you and your husband have decided on a name, would you mind coming down to the office and registering him? Thank you.'

After Joe's birth in SE5, a typed slip was sent to me at home, telling me, 'Please register your child without fail by such-and-such a date.' So I went off, between feeds, to the Nissen hut beside the hospital, into an office full of tables, where I stood in a queue. I sat down at the next available table, and the registrar asked, 'When was baby born?'

'New Year's Day,' I said proudly.

'Oh, yes,' she said. 'That would be January the . . .?'

'First.'

'Are you married to baby's father?'

'Oh, yes.'

'What are you going to call baby?'

'Joseph John Lancaster.'

'And it's a little . . .?'

'Boy.'

(Both these encounters were on the NHS, just a different postcode.)

Joe's arrival meant that we no longer fitted into our little terraced house in Barnes, and needed to move somewhere altogether bigger.

I'd just done two years with the Royal Shakespeare Company, but, as they didn't hold any plans for me in the future, I was now on the lookout for something else. Pru knew the director Toby Robertson, who ran the Prospect Theatre Company, and we went round to see him at his house in Wandsworth.

I ended up working with him for seventeen years, on and off, but in the meantime we'd seen a house just round the corner from him that we rather liked, so we took out a mortgage and bought it.

We've been living there now for forty-five years, very happily: it's well served by buses, and but a short walk to wonderful Clapham Junction, perhaps the most useful

A gaggle of cherubic children are all very well and good, but they are famously inconsiderate when it comes to taking up real estate in the house.

railway station in Europe. I don't like driving in London, as I can never find anywhere to park; and so Pru and I use public transport all the time. For one thing, now we have our sublime Freedom Passes, we can travel without cost on buses, the Underground and far beyond the suburbs by National Rail. Sometimes, if one of us has recently hit the screens, people say they're surprised to see us on the Tube. 'I thought you people went around in limousines?' they ask, in tones of disappointment. Well, I suppose some do; but we think it's part of our job as actors to study the people we see in daily life, to try to imagine what they do, how they live, what gets them out of bed. You can't do that sitting in the back of a taxi.

NARROWBOAT
NOVICES

ALL THIS MAY SEEM a long way from the world of canals. However, in 1976, our friend Lynn Farleigh rang and offered to lend us her narrowboat for a fortnight. I'd rather forgotten about our domestic waterways, and for Pru it was quite a new subject. She knew about the Suez Canal, and the Panama Canal, and, yes, all right, the Manchester Ship Canal, but the idea of going from Oxford to Banbury by water seemed to her pretty remarkable.

In fact it turned out to be perhaps the best holiday of our lives. The summer of that year remains famous for its uninterrupted warm sun and cloudless skies; the swans, ducks, moorhens and water-voles, the herons and kingfishers were enjoying life to the full as the canal wound its sinuous way through the bewitching Oxfordshire countryside. The boys, then twelve and ten, got so blissfully exhausted operating the locks and lift bridges that by six o'clock they were flat on the their bunks, and Pru and I could open a

A cunning ploy to get children into bed early is to recruit them into intensive manual labour. Never fails.

bottle of wine and sit watching the setting sun and hearing the gentle trickle of water from a remote lock gate.

For us it was the start of a lifetime's fascination. I don't know how a pair of busy actors found the time, but somehow we determined to explore as much as we could of the whole canal network. This took us years, of course. We had to fit it in between jobs: picking the boat up from wherever we'd left it, and then abandoning it somewhere else, perhaps knocking at a nearby cottage and leaving our phone number just in case there was a problem.

Lynn eventually sold us her boat, which went on to serve us for several happy years until we got tired of having to dismantle the dining table every night to turn it into a double bed. We would delay this irritable task for so long that often we would wake up at three in the morning, our heads on the table, shivering with cold. So, instead, we commissioned our own, longer, narrowboat, bought the hull, designed it and had it fitted out at Tooley's famous boatyard by the marine engineer Barrie Morse. The launch took place at Banbury in the winter of 1988: we christened her after our brilliant female accountant who had managed somehow to allow the purchase against tax ('Extra Office Space').

Incidentally, by giving our boat a feminine identity we are actually flying in the face of canal tradition. To the working boatmen of old, their craft might proudly display an elegant female name, but was still firmly referred to as 'it'. Similarly, 'port and starboard' were frowned on: they

We had no idea, at the time, how many happy memories from our lives would come to pass on our precious boat.

preferred left and right; also a boat had a front, and a back. I'm afraid Pru and I have found it impossible to adjust; I suppose secretly we like to pretend we're accomplished global seafarers.

We kept, and still keep, a logbook of all voyages. Somehow the actual launch didn't get logged (too much champagne, perhaps) but here is the entry for the boat's first family trip (my father, aunt Joyce with her new dog Sally, Pru and myself), from Banbury up to Cropredy and back:

Sunday 18th December 1988: Everyone very impressed with the new boat, the cooking stove and to a slightly lesser extent, the loo. T still discovering that a 60' narrowboat is at least 12' harder to navigate than a 48 foot one. Going round sharp bends you will have your rudder aground if you're not careful. Well, there's not much water in the Oxford Canal just now. Chief delight is how *warm* the boat gets, and how quickly. It was well nightfall by the time we got back to Tooley's, and found there was no mooring available: while we were away, Barrie's new client had crept into our spot. We moored alongside, but it was clear that our senior passengers couldn't manage the plank, so we moved up to one of the staithes [landing stages] and put them ashore. It was then that the stern line got wound round the prop shaft, and I spent the best part of an hour on my face, with an arm in dark freezing water to unravel it. Never mind – I'm pleased to learn it *can* be done.

On our earlier boat we had already covered a great deal of the national waterway system. We'd sailed up as far as Ripon in North Yorkshire, the furthest in point in England to be reached by canal; we had explored the intricate water-webs of Birmingham and the Black Country, and taken the Grand Union (by far the prettiest way to view Milton Keynes, if you have to) all the way back to London.

One important route, though, was closed to us as yet. Since the afternoon I saw that farmer fall into the mud, the Kennet and Avon had undergone a spectacular reversal of fortune. A petition with 20,000 signatures had been addressed to the Queen, and it led to a phenomenal restoration programme carried out by volunteers who included servicemen, prisoners, students, schoolchildren and Boy Scouts. The entire eighty-seven-mile length across the country from Reading on the Thames to the River Avon at Bath, and thence to Bristol, was reopened in 1990; and it occurred to the television company HTV West that it might be nice to arrange for Pru and me, who were quite well known to the area, to be recorded on film as the very first boaters to travel the reclaimed waterway.

We got two friends to accompany us on the journey, and set out on our adventure, beginning at the start of the actual canal in Bath, just below Pulteney Bridge. Nervous but enthusiastic members of the K&A Trust gave us a send-off and followed our progress at a discreet distance, while a considerable crowd turned out to watch us negotiate the momentous sixteen-lock staircase at Caen Hill.

At one point we had to break our journey for a few days while the stretch of canal ahead of us had yet to be supplied with water; but, apart from that, the only interruption to our

Our boat's stately procession, via a flatbed truck, several hundred yards down the canal past the top lock, so that Her Majesty could be the first to sail through the new lock gate. One suspects she probably wouldn't have minded us going first.

maiden voyage was a readjustment to allow the Queen to perform the official reopening ceremony in a week or two's time. Her Majesty's ceremonial barge had to be the *first* to pass down through the top lock at Devizes, on its way to the spot where the reception was to take place. A journey of only a few hundred yards all told, but necessitating our sixteen-ton narrowboat to be craned out of the water, loaded onto a flatbed truck and driven round the streets of Devizes before being winched back into the canal a few yards further on. So, when we boast of having sailed the whole eighty-seven miles of the canal, it would be more accurate to say eighty-six and three-quarters.

Once we'd watched the perturbing spectacle of our boat being dropped back into the water, we were able to attend the royal ceremony, and talked briefly to Her Majesty, who seemed pleased with the event and remarked on the distinctive variety of butterflies that she'd witnessed on her short trip.

HTV's four thirty-minute episodes were much enjoyed when they were shown locally, and I think may well have drummed up tourist interest for the canal, which at that time it rather needed. Now, of course, with its beautiful scenery, remarkable features of engineering and pleasant, welcoming towns and villages, it's top of the location list for many canal enthusiasts. It certainly found a place in our own lives: for years we kept our boat at Newbury, performed shows at the little theatre on Devizes Wharf, 'adopted' a lock at Combe, and became Patrons of the K&A Trust.

Wonderful though such memories of canal life are, being actors also meant travel of a different nature, as we'll see.

Stalin in *Master Class*, at the Old Vic in 1983. We didn't take this show to Communist Yugoslavia.

TOUR DE FARCE

OVERSEAS TOURING, particularly of classical drama, meant that we were working abroad quite a lot. We wanted the boys to be with us as much as possible, so the current au pair/nanny/mother's help did get to go on some quite nice holidays: in Australia, the United States, Europe and the Middle East.

There's one particular tour I remember: through Yugoslavia (as it still was then), Turkey, Jordan, and Egypt. I was acting as staff director for our three productions: *Hamlet*, *Antony and Cleopatra* and *War Music*, a verse/music/dance staging of Christopher Logue's translation of the *Iliad*; and it was my job, as well as playing Claudius, Enobarbus and the Storyteller in *War Music*, to fit all three shows into the various venues that we'd been allotted.

We opened in Istanbul, in the open courtyard of Rumelihisarı Castle on the shore of the moonlit Bosporus, which served very excitingly for *War Music*; but the next night's performance was *Hamlet*, and somehow we hadn't taken into account the presence of a nearby mosque, from which the Muezzin's loud appeal to prayer came just as our Marcellus observed that something was rotten in the state of Denmark.

Although afterwards the tour was accounted very successful by audiences in all the places we visited, we did seem to be dogged by various kinds of misfortune. Our next date was Ljubljana, in Slovenia; and when we arrived we found that a number of our wardrobe skips had been lost in transit. Maybe our swords and shields had been impounded for safety by customs officers who were gaily trying on our wigs and frocks. They eventually did arrive,

The views along the Bosphorus where we staged our production of *Hamlet*, are spectacular. © *Getty*

but too late, and that night the forces of Rome and Egypt had to go to war unwigged and unarmed.

Dubrovnik was next, and for *Antony* a special arena had been built in the square in front of the Ducal Palace. We could not begin the performance until after dark, when all the cafés had closed and all the starlings that inhabited the square had gone to bed for the night. And then the stage lights came on, and a thousand birds awoke to greet the supposed dawn, lustily drowning out every word that was spoken on stage (classical actors didn't use microphones in those days!). Not many people stayed after the interval.

We moved on up the coast to the Roman port of Split, and here we performed on the forecourt of the splendid Palace of Diocletian. It looked wonderful; the only worry was that, if an actor had to make an exit on one side of the stage, and re-enter shortly afterwards on the other, he or she had to go down some steps to a courtyard and along a narrow public alleyway into a side street, and thence back into the palace by a side entrance. Sometimes this had to be done fairly rapidly.

One night, one of Cleopatra's court, wearing a very bulky robe, had got as far as the alleyway when he found his path blocked by a very fat local citizen coming in the opposite direction. 'Back up! Back up!' shouted our actor, but the man, muttering his rights, stood his ground. Finally the resourceful actor drew his sword and brandished it above his head, and at that his opponent turned and ran, screaming.

Pru and the boys came out to join me in Jordan, where we were to play in the Palace of Culture in Amman (actually a 6,000-seat basketball stadium). The stage lighting was completely inadequate; I complained, and was promised that extra lamps would be delivered and installed in good time. They weren't. I had not then learned that in some Arab countries it is considered discourteous to refuse a request: they say yes, and then don't do it. I was prepared that this State Performance of *Hamlet*, in front of King Hussein and a possible audience of 6,000, would have to happen more or less in the dark. In the event, only about 350 people turned up, and we seated them together in a block and directed what lights there were onto what they could see.

The security arrangements for the royal visit were exhaustive. Our set, together with the furniture and properties, was closely examined; even the Leichner sticks of makeup still used by some of the older actors were probed by sharp knives in case they disguised cartridges. During the performance, a dozen heavily armed soldiers were stationed backstage, one of whom was discovered when Hamlet draws back the arras in the bedroom scene, gazing down in concern at the dead Polonius.

We were in Amman for quite a few days, and, while we were there, the company fell prey to various gastric disorders – though mercifully our own family were spared. At one point eight actors and two musicians had been committed to hospital, a couple of whom dragged themselves out each night for the performance and quickly returned to their hospital beds afterwards. Everybody had to cover for somebody else, we combined parts, cut whole scenes, and just about managed to make sense of the plot.

The Theatre of the Sphinx in Cairo, our final port of call, has a long stage raised four feet above the desert sand, and at each end of the stage is a short flight of steps, in the dark, down to the dressing-tents. On our closing night someone had moved these steps; I didn't know, and fell four feet. Not a long fall, but unexpected, and I tore my Achilles tendon. (Later, it broke completely, which caused me a lot of bother and I had to have an operation to fix it.)

That was the end of an eventful tour, but the company remained in fine spirits; and those of us who were sufficiently recovered from our ills to hire a pony first thing in the morning and ride down the long valley to Petra – 'a rose-red city, half as old as time', as John William Burgon

No matter how ridiculous and cumbersome the costume, no matter how un-ideal the location – no matter how temperamental the horse – the show must go on.
© Rex Features/ITV

put it in his magnificent poem, also called *Petra* – and see the incredible Roman Treasury carved into the rock, enjoyed an experience that made up for all the trials and tribulations we'd encountered during our wanderings.

While I love travel, Pru maintains she doesn't really like opening the front door to put the milk bottles out. It's not true: she loves it really. I think.

The boys like it, anyway; at least Joe did: 'For one half-term, Ma took us to Paris, Strasbourg and Luxembourg and we stayed in cheap, dusty, elegant European hotels. I really liked seeing different places. We went to Jordan, Egypt and the US, and to Australia twice, which I loved.'

Pru and I have worked in Australia on numerous occasions, and at one time we even thought of going back to live there. But then we wondered whether we would still be welcome: Australian acting was now doing very well, thank you, without any assistance from the Poms!

Our favourite place in Australia was Perth, technically the most isolated city on the globe, 2,300 miles away from its nearest neighbour, Adelaide, across the desert of the Nullarbor Plain.

Because of its distance from anywhere, it has a vibrant social life of its own, and an appetite for travel that is external rather than internal. Someone will tell you they've made their annual pilgrimage to Bayreuth, Salzburg and the Edinburgh Festival; but poor Auntie Alice in Sydney has had to go unvisited.

We appeared a number of times at the Perth Festival, counted among the best literary and musical events in the world. Pru directed *Uncle Vanya* for the Western Australian

This photograph was actually taken in Melbourne. You can see from Pru's outfit that, perhaps, we had been spending a little too much time in Australia. . .

Theatre Company, with me as Vanya (I like being directed by Pru, she's *very* good). I did a term with the University of WA, as Director-in-Residence, and did a student production of Middleton's *Women Beware Women*.

The only trouble with Perth is that its amazing lifestyle can prohibit you from working. The city is built round a beautiful harbour, too shallow to permit commercial craft, so nearly everybody has a little boat of some kind, in which, armed with a pack of Foster's, they can sail, row or motor down the River Swan to Fremantle for a seafood lunch.

Pru's remark about her reluctance to put out the milk bottles is, I'm afraid, utter rubbish. For twenty-five years – yes, twenty-five – she has been doing single performances of *An Evening with Queen Victoria*, a programme compiled from the Queen's own diaries, and performed with the tenor Ian Partridge and the pianist Richard Burnett. She plays Victoria from the age of sixteen until her death, without any change of makeup, and the effect is remarkable. The three of them have taken the show all over the world, and I can't think how many air miles they must have clocked up. So forget the milk bottles.

THE BIRTH
OF *GREAT CANAL*
JOURNEYS

IT'S AUTUMN, getting cold, and it's time our boat went into dry dock to get its bottom blacked. If you own a canal boat, and you're not intending actually to live on it full-time, of course it's essential to find a winter home for it where it will be safe, accessible and looked after. For years we found such a place at a boatyard in Newbury owned by Bill Fisher and his wife, who were pillars of the Kennet and Avon Canal Trust.

I can't remember why, after many years, we decided to relocate: I think perhaps we just wanted some new scenery. Anyway, we found a mooring at Banbury, not far from where our boat was actually built. The mooring was directly on the canal, but to reach it by land you had to go through a locked gate (we frequently mislaid the key and had to ring the owners at home at all hours: they didn't live on the premises). It was only when they decided to retire and sell the mooring that we thought maybe the time had come to move on.

Various canal obligations had brought me into contact with a man called Tim Coghlan. Tim is king of the canal network's boatyard proprietors and held sway at his vast marina at Braunston, Northants. Situated as Braunston is, at the hub of commercial canal traffic, stretching north to Coventry, northeast to Leicester, southeast to London, southwest to Oxford and due west to Birmingham, it has always been a national centre for mooring, boatbuilding and repairing, and a gathering of the canal community. Here we made our resting-place.

People who are unused to canal boating come to us sometimes for advice. How can you can ever be comfortable in the confined space of a narrowboat? they ask. Is there any discipline that one has to learn, to be able to cope with dimensions of (in our case) 60 foot by 6 foot 9 inches?

Well, yes, you have to learn to be quite *tidy*; but in a well-designed boat there should be plenty of storage space for books, crockery, kitchen utensils, glasses, bottles, tools, bedding and cleaning materials – just don't leave things about.

As for human beings: our boat can actually accommodate six – two in a double bed and four in individual bunks – but people have to know each other pretty well, and not snore.

If you're just hiring a boat for a week or two, resist the temptation to bring loads of exotic groceries, toys, CDs and board games: just take what you need to arm you against possible continuous rain (it can happen). You need your energy, so eat up. Pru enjoys cooking on the boat (actually rather more than she does at home). Take advantage of the various canalside hostelries that offer good food, local real ale and usually entertaining company.

You shouldn't set too much store by getting to a particular place by a particular time. As one of the ladies we met while filming in Scotland put it, 'If you want to make God smile, tell him your plans.' There will always be a church you'd like to peer into, a hill you would like to climb, a wood you would like to explore; and you just feel you haven't the time.

The Nicholson waterways guides are essential. There are seven of them, for different areas of the country. They describe the terrain in front of you, map out your route, give you distances, show aqueducts, cuttings and tunnels and identify the locks, bridges, moorings, boatyards, water and sewage points, and pubs. They show you the off-canal availability of shops, post offices (sadly, not many), bus stops and railway stations.

Moorings, these days, can be difficult to find. On some of the most popular canals you can go past perhaps a mile of tethered craft before you can find a space. If you're leaving your boat to go ashore, lock up firmly. It is rare to be burgled; though our boat was once broken into, for, curiously, a packet of muesli and some corduroy trousers. It turned out that the perpetrator was in fact known locally: he made a practice of stealing from moored boats during the summer months, but, as the weather grew more unkind he would arrange to be apprehended, and spend the winter as a guest of Her Majesty. Then he'd come out and start all over again.

Rocketing housing costs, particularly in London, mean that people are flocking to live in narrowboats and one can see the resulting congestion for oneself, as here on Regent's Canal.
© Getty

More and more people, especially in cities, are buying a second-hand narrowboat to live on, at perhaps a tenth of the price they'd be asked to pay for a one-room flat. But that's only the start of it: there's a licence and insurance; and mooring fees can sometimes be really expensive in a marina or secure boatyard. If you can find a space along the towpath to moor up, that's all right for a fortnight, and then you'll have to move on somewhere else; that's the law.

Do we live on our boat for any length of time? we're asked. Yes, often, when there is a canal with a secure mooring near the theatre at which we're playing. It's a home. We've lived aboard in Bristol, in Bath many times, and in Leeds for a while. Sam borrowed the boat for two seasons in Stratford-on-Avon.

Pru regards the boat as a second home; the only thing she really misses is the garden. Pru is a very keen gardener. We have a lovely garden at home, and when she's away she likes me to attend to it, and report on progress. I'm a novice. This is a letter I wrote to her in 1968:

The Garden. Ah, well now. Umm. Yes. *Yes.* Right you are. Well, here goes then. RIGHT. From the beginning. First things first. My general impression is that there are more leaves than flowers. Just so. Quite a lot more, in fact. The next thing that strikes the intelligent observer is that what flowers there are, are to be found on the left, as you look out of the window. None on the right. No. What *are* these flowers, you will wish to ask? Well, they are mostly those pink ones, you know, several to a stem, also available in blue. And white too. Oh, and purple. Or are those different? Anyway, there are about ten in all.

Now I don't want to stick my neck out here, but I believe I may be right in saying there are some daffodils on the same side of the garden, and one thing I'm pretty sure of is that there were a lot more of them at this time last year. At the end of the garden we seem to have a lot of tall green stuff with small yellow bits on top. A considerable amount. Probably more than anyone else in the road.

Well, that's about it. Yes, I believe that just about

wraps it up. Anything else you want to know about the garden, you know you have only to ask.

Some people like to create gardens on the tops of their boats. If you're living aboard permanently and not moving, that's fine, but, if I had to steer while trying to peer between hollyhocks and sunflowers, I think I might bang into things. (I do bang into things sometimes, and not always by accident: Channel 4 rather like me to have a bump from time to time.)

We were beginning to be recognised on the canal circuit, and I was approached by Carlton Television in Birmingham to introduce a series of weekly half-hour programmes

An early precursor to Great Canal Journeys.

about canals and their history, entitled *Waterworld*, dealing mainly, but not solely, with the West Midlands. It was put together, produced and directed by the admirable Keith Wootton, and there was some splendid footage and lovely archive photography. All I had to do was a short piece to camera to introduce each episode. We did it for nine years, and it was very popular in the Midlands. We were preparing for a tenth year when Carlton told us that, due to cuts, locally shown programmes had become uneconomical. So why not show them nationally, on the network? we asked. Ah, no, they explained; you see, there are lots of areas in the UK that just don't *have* canals. We said, well there are lots of areas in the UK that don't have volcanoes, or glaciers, or rainforests, but they make pretty good television. But, no, they wouldn't budge.

I don't exactly know how *Great Canal Journeys* came about. Perhaps someone at Channel 4 had seen *Waterworld* and thought it was a good topic that could be expanded. Or did the idea come fresh from the independent company Spun Gold Television, and did they pitch it to the Channel? I don't know, and it doesn't matter. Spun Gold are a pleasure to work for, and C4 are immensely supportive.

The unit that we work with is very small: there is Mike Taylor, who is in overall charge, functioning as producer, director and screenwriter; James Clarke was our senior cameraman on all the early episodes, with Gary Parkhurst on second camera; Sam Matthewson did the sound, and we had different production assistants and a different runner for each shoot. As time went on, the wonderful Catriona ('Trina') Lear joined us as PA, researcher, caterer, wardrobe mistress and line producer, and Pru and I made up the small, happy and efficient family. In the last couple of years we were joined by my darling daughter Juliet, who, after her mother and I parted, came to live with Pru and me, and eventually occupied our basement flat. Juliet is a professional hairdresser, and, as our unit wasn't grand enough to boast a dresser or a makeup artist, Pru fell on her neck sobbing with gratitude.

Making a programme that is meant to look spontaneous and impromptu does of course take quite a lot of planning. Preproduction work starts with selecting a location, somewhere that we, and we hope the audience, will be

Whilst Pru and I are front and centre of the show, none of it would be possible without our fabulous – and very patient – crew.

excited by. What will the weather be like? Might there be any political problems leading to restricted access? Where exactly should we go? What should we see? Whom might we talk to about the history of the place, and of the canal or river? What is there of literary, artistic or musical interest? And so on. Also, we have to charter a boat.

When we feel we have enough potential ingredients to make a good programme, a route has to be planned that if possible can include it all within a six- or seven-day shoot. Mike, with Trina and the leading cameraman, will first of all go out there on reconnaissance to plan how to make it all possible and think how to capture it scenically.

Meanwhile, Pru and I will do a bit of reading: is there a local writer, a poet or dramatist that we should be able to quote; any local legends we should know about? Latterly, we have had the services of a researcher to do a lot of the groundwork, and they have often come up with some quite surprising things.

Something we had to decide upon, before we started making the series, was how open we should be about a condition in darling Pru that had been diagnosed as vascular dementia. In appearance, this is very similar to Alzheimer's disease, but develops more slowly (at least it has in her case).

It must have been at least twenty years ago that I came to see Pru in a play she was doing at Greenwich, and I thought, There's something wrong here. It wasn't that she was uncertain about her lines – she obviously knew them perfectly well – but there was just this sensation that she had to *think* for a millisecond before she spoke.

Now that's not how Pru works: throw her the ball, and she's caught it and thrown it back to you in a trice. So after the show I just asked her if she was feeling OK, and she said yes, perfectly, thank you, what do you mean? I can't remember how I persuaded her to go and see a specialist, who sent her for a brain scan, and they made the diagnosis.

Now, as I say, the progress of the condition was very slow, and still is, but before long it was beginning to be apparent to friends and to employers: things like repeating a story three or four times in half an hour, forgetting some vital thing she had been told, and asking the question again (Judi Dench, in her brilliant performance as Iris Murdoch in *Iris*, portrayed it exactly).

I had to think: is this going to be apparent when we're filming; or, even if it isn't, how many people will have heard rumours and be watching out for indications? So I decided I couldn't possibly pretend to ignore it: to do so would be foolish, dishonest and altogether wrong. With Pru's agreement and that of our TV producers, I made a clear statement about it on screen, and revised this as we went along.

How is it for me? Well, painful of course when I remember the wonderful girl I fell in love with, so clever, so funny, so adventurous. But one mustn't do that, mustn't dig into the past. Live for the present, just take it day by day. Day by day I don't notice that much difference in her: physically she's in amazing shape for her age, and still likes going to the theatre, to concerts, restaurants and so on, and especially being on the boat. She's well aware of the problem, doesn't like talking about it, and is not letting it get her down. We really do have a lot to be thankful for.

PART 2

THE KENNET
AND AVON

ONCE WE HAD DECIDED we were in business with the series, it became a question of choosing four different, and contrasting, canals, to make up a group of four sixty-minute programmes.

It was natural, I think, to choose for our first subject the dear old K&A – the Kennet and Avon. Pru and I knew it well it was easily accessible, very attractive and now immensely popular. Travelling along it now was a totally different experience from how we had found things in 1990, when we had been the first boat to venture on the cut for forty-two years.

In those days, the few people strolling along the towpath would glance at us curiously, intrigued by the unfamiliar sight of a boat. Now there were flocks of sightseers following us up the Caen Hill Flight, picnicking on the towpath; leaning over the parapet of the Avoncliff Viaduct, where one can celebrate the combined genius of Rennie,

Brunel, Macadam and God as canal, railway, road and river merge together.

The attractive canalside café at Bradford-on-Avon was crowded, and the proprietor remembered us from twenty-three years ago. In those days he had no more than a little booth selling tea and ice cream; now it had fully fledged restaurant proportions, all this owing to the regeneration of the canal.

Further along, a lady had opened a hairdressing narrowboat, and Pru booked in for a cut and style.

We went further on to Seend Cleeve. Still intact after fifty years (they are literally indestructible) are those concrete bunkers guarding the canal from World War Two invaders. How were these people supposed to arrive? I wonder. By canal, in submarines cunningly disguised as narrowboats? The crew apparelled as nuns? (As children we'd always been taught this was how the fiendish German spies routinely disported themselves.) Could that boat, painted 'George and Edith. Dortmund', perhaps arouse the suspicion of Captain Mainwaring and the Home Guard? Well, they never came, but the bunkers are still there just in case.

We'd decided for the purposes of the series to restrict ourselves to the western section of the K&A, from Bath to Devizes. If you continue thence eastwards you get to Pewsey, Hungerford and Newbury, and from there on

Avoncliff Viaduct: a stunning marriage of nature and engineering. © Getty

it gradually becomes more urbanised until at last you get to Reading, and the Thames. Most of this last section actually uses the existing River Kennet and, when attempts were made by the government in the fifties to close that part of the canal, it was pointed out that there existed no statutory powers actually to close a *river*.

This is my part of the country, really. I've spent much of my life in Bristol and Bath, in the Mendips and on the Bristol Channel coast. At the end of the war I made many train journeys backwards and forwards from our flat in Bristol to see my father, who was working in London. I loved the Great Western, and used to lean out of the window (you could, then) and watch the stations flashing by: Wantage Road, Challow, Uffington, Shrivenham, Stratton Park Halt... Lovely wayside stations, all gone now, thanks to Ernest Marples and his sidekick Dr Richard Beeching (transport minister and British Railways chairman respectively, who closed so many rail routes in the 1960s). 'Who needs these *local* stations?' they asked. 'Soon everyone will have a motor car!'

Yes.

We did some provisionary shopping in Bath, and got aboard. We were very pleased with the boat we were given to use. It would not have been practical to fetch our own, and we had been very well served here with a craft that was well designed, easy to manage and good to cook on; also, the beds were very comfortable, so no need for hotels – Pru and I slept on board every night.

We set off. It was the August Bank Holiday, and there were a lot of other boaters using the canal, some of them not very expertly. A lot of the time James knelt on the roof, filming me, while Mike sat beside him firing me questions. Thinking of things to say, while peering between the two of them to see where I was going and avoid the other boats was sometimes not easy, but we got into a rhythm.

I was sometimes criticised by onlookers for staying by the tiller and letting Pru do all the work; but, particularly in broad locks, it is essential to have someone in control of the boat, particularly if you're 'locking down'. There is a real danger, if you float too far back against the top gates, of your rudder getting caught on the cill of the gate; a situation that will not be apparent until it's too late and your stern is left hanging there, while the bow continues to drop downwards as the lock empties.

If you discover what's happened in time to close the bottom paddles and quickly open the top ones to refill the lock, you may escape with nothing worse than a bent rudder (though that's nothing to laugh about); but serious, and occasionally fatal, accidents have happened through getting 'cilled'.

We had to try to keep in the central channel of the canal, because it hadn't rained for a time, and some places were very shallow. It's remarkable that the whole length has to be fed by a single reservoir, Wilton Water, that sits on a canal summit only three miles long; a situation that must have come about through the meanness of investors refusing to pay for the extra locks or embankments that would have allowed a longer stretch of water. As it is, periods of drought can pose a very serious problem, and, while there is now an electric system of supplying water from Wilton, valuable assistance is still supplied by Crofton Pumping Station.

This historic piece of engineering is a long way east of Devizes, and therefore was not really on our route; but we felt we had to go and have a look. The old building contains two Cornish beam engines, the older built by the early nineteenth-century firm of Boulton and Watt; while the other bears the plate 'Harveys of Hale, 1845'. Between the two of them, two tons of water can be lifted up to the summit at every stroke. It is wonderful to watch this happening.

Attitudes towards the reopening of the canal differ somewhat. Soon after the work was completed, I had been to a wedding just outside Bath, and I was approached by a very smart lady in a white hat, who asked me, 'Didn't you write something about canals in a newspaper somewhere?' She was quite sharp with me.

I admitted that I had.

'Well, I *live* on the canal,' she said. So I asked her where.

'In Avoncliffe,' she said. 'Bradford-on-Avon.'

'Oh, yes, I know it,' I said. 'It's very pretty.'

She came right up to me. 'It *was* . . .!' she admonished me. 'It was *very* pretty; and then they went and put *water* in it, and now I have to watch a lot of boats going by.'

I apologised and went away to get a drink, but she kept following me with a resentful gaze. It was clear she held me directly responsible.

Bradford-on-Avon is a lovely place, if you're prepared to forgive the presence of water. There's an attractive lock,

Pru was uneasy about handling the boat in broad locks; they do swing about a bit and bang into the walls if you're not very careful and not very strong; so she preferred to be the one to get off and wind the paddles, and open and close the gates. As time went on, I began to see this was getting a bit tough for her, but she was happy doing it; and, of course, if a paddle was too stiff, Mike or one of the crew, off-camera, would come and help.

55

at one end of which is a small domed building: the town lock-up, in which wrongdoers were made to pass the night before being presented to the magistrate in the morning. The Town Bridge was built in Norman times; other buildings date from the flourishing woollen textile industry here in the seventeenth century.

The town was very busy as we passed through, and boats were coming in and out of the popular Hilperton Marina; but gradually life quietened down, and we cruised peacefully along the five-mile lock-free pound to Semington. It was idyllic: life at four miles an hour, beautiful woods on one side, the rolling Mendips on the other, the gentle swish of the water. We looked at each other and thought, Yes, this is the way to live.

The canal keeps up a long-distance correspondence with the River Avon: it reappears from time to time until it goes away to seek its source in the hills near Malmesbury. We got to Semington village. Here to the north we could see the junction with the Wilts and Berks Canal, built in 1810 as an attempt to link up with the Thames at Abingdon. It was a short-lived undertaking, with plans for its abandonment already taking shape as early as 1874.
The collapse of an aqueduct near Chippenham in 1901 sealed its fate, although there is energetic popular interest now in restoring it: a few years ago the Duchess of Cornwall symbolically cut the first sod of a proposed new course for a revitalised Wilts and Berks. So who knows?

As we approached the Foxhanger Locks at the foothills of the Caen Hill climb, an element of unlooked-for drama occurred when we were told that someone had crashed into a lock gate and pulled it off its hinges, and consequently all traffic upstream was suspended in both directions. This was very bad news, as the whole Caen Hill experience would be denied to us – but, fortunately, the brilliant engineers of the K&A Trust got the thing back into service again in time for us to sail up this astonishing world-famous staircase.

We reached the top, and were able to join the revellers at Devizes Wharf for their festival knees-up. As well as the mandatory display of morris dancing (a spectacle that I'm afraid strains my attention quite early on), we had a troupe of ladies and gentlemen of a certain age performing nineteenth-century courtly dance, very charmingly. There was a magician, a number of solo musicians, and a great deal of very good beer. It was a deliciously warm evening, with a lovely sunset; and it was good to see so many people enjoying themselves on the bank of the canal that they had lost for so long and was then restored to them.

The spectacular – but daunting – sight of Caen Hill's sixteen lock gates. The flight is beautifully designed with side-ponds between locks, so that boats having cleared a lock on the way up will turn in to the pond, leaving down-boats to come straight into the vacated lock; saving both time and an incredible volume of water. © Getty

THE
ROCHDALE

IT HAD BEEN DECIDED to do all four episodes pretty well back-to-back: this had taken a lot of planning, and at the end would mean a very long period of editing, but for Pru and me it was good that by concentrating all the filming over a few weeks we weren't blocking ourselves off from other work.

Curiously enough, before embarking on Episode Two – the Rochdale Canal – I had been asked to do a few days on the TV serial *Last Tango in Halifax*, covering at least part of the same geographical area. I was to play Derek Jacobi's lost brother Ted, and, having been his stepfather Claudius in *Hamlet*, and his gay hairdressing partner in *Staircase*, I was very keen to do it. It worked out very well, and I ended up in the same pub in Hebden Bridge, drinking with the *Last Tango* cast on Monday, and on Tuesday with our camera crew for *Great Canals*.

The Rochdale Canal could hardly be more different from the K&A. It's the first of three canals built two hundred years ago to provide a trade route across the Pennines, and is a very steep climb.

It was really very hard work: looking back at a DVD of the episode the other day, I was really astonished to see the amount of effort required to cover each section of the canal we'd planned to travel in a day. The locks were tough. It was nearly always Pru's job (her preference) to open the paddles and handle the gates, and I am speechless with admiration, not unmixed with a sense of guilt at letting her do it unaided. She never complained; we worked together as a crew, and we both had our jobs to do.

Pru and I have been doing this for a while, and make a pretty good team, when all's said and done.

After we got to meet our hire boat at Sowerby Bridge,

60

and said hello to the proprietors, we thought we'd make a detour and pay a visit to one of the other trans-Pennine contenders in Calderdale, the nearby Huddersfield Narrow. Pru and I knew this canal pretty well, as we'd been around when its restoration programme was coming to a climax in the nineties. But, at that time, the reopening of the three-and-a-quarter-mile-long Standedge Tunnel, which was needed to complete the route of the canal, seemed still a long way off.

The original challenge of broaching the Pennines to connect the Yorkshire Mills with the Lancashire market and beyond was taken up by the three different canal companies in three different ways. The Leeds and Liverpool opted to go round the hurdle, the Rochdale went over the top of it, and the Huddersfield Narrow ploughed through the middle.

The Standedge Tunnel took seventeen years of digging, through the hardest kind of rock; an incredible feat of engineering costing many lives. It survived in its original state until the canal closed in 1944, but finally was triumphantly reopened in 2001. Now we came to have a look, and with the guidance of Fred Carter, the tunnel expert, I steered our boat through. It was an extraordinary experience, taking nearly three hours, and Pru was genuinely unable to rid herself of the sensation that she might never come out the other end.

Just as I had felt a bonding with the Somerset countryside, the banks and dales of West Yorkshire found a resonance with Pru. I don't quite know why; she had an uncle up here who ran a mill in Huddersfield, but hadn't really spent any

more time up here than she had in Bristol. She just liked the people, and how they spoke, and how they lived their lives.

More recently of course we've both worked a great deal within the county: in York itself, Leeds (we're Patrons of the West Yorkshire Playhouse and have performed there several times) and Bradford in the west and Sheffield in the south. I've paid many visits up to the old Georgian Playhouse in Richmond (North Yorks), and I spent early time as an assistant stage manager over in Hull.

Also, I was actually born in Bradford.

Pru, however, wouldn't have it that I was a genuine Yorkshireman. She reminded me that I happened to be born there only because that was where my father was working, on tour, at the time. A month later, it would have been Eastbourne. She has a point.

We got back to Sowerby Bridge and picked up our boat to travel the Rochdale. It's a Canal that tells a story. For most of the way, it passes between handsome old mills and textile factories that had once been part of a booming economy, but that, in the 1970s, had come to realise that India provided cheap labour and easy access to the cotton fields; so most of these majestic buildings were now turned into flats or housed small businesses.

Our son Sam came up to join us – he has always gone along with our love of canals, and he and his partner Laura have often borrowed our boat. Soon they were about to have their first child, though, so it was likely to be another four or five years before they would risk it *en famille*.

We passed Luddenden Foot, where there used to be a railway station, at which Branwell Brontë was briefly

Opposite: The light at the end of the tunnel: The Standedge tunnel took seventeen years to make. Men with torches and pickaxes dug through three and a quarter miles of hard rock, sometimes at a rate of as little as three yards a week. Men died to complete it and travelling through is claustrophobic in the extreme – with the knowledge that over 600 feet of rock weigh heavy over one's head.

An Austerity Class steam locomotive pulling a coal train travels through Luddenden Foot. © *Getty*

employed as a clerk, and was sacked for the usual reason of drinking on the job. Then we stopped at Mytholmroyd, home of the poet Ted Hughes, met Nick Wilding, a Hughes literary authority; and we read a couple of the poems. One of them, 'The Long Tunnel Ceiling', was written under Halifax Road Bridge, where we tied up, and is about the shock of hearing a loose brick falling from the roof of the bridge; here's a short extract:

Suddenly a crash!
The long gleam-ponderous watery echo shattered.

And at last it had begun!
That could only have been a brick from the ceiling!
The bridge was starting to collapse!

But the canal swallowed its scare,
The heavy mirror reglassed itself,
And the black arch gazed up at the black arch.

We went through what is arguably the deepest lock in our canal system. It's a broad lock, too, of course, and it's difficult to control the boat as the water rushes in. Three concerned-looking Canal & River Trust (CRT) lock-keepers looked down on us from a great height, but we made it.

Another of Ted Hughes's local canal haunts was the pub at Stubbing's Wharf. He was seduced by the gloom, the seediness and depression, but the place has obviously been taken over and the atmosphere, if you forget about the wallpaper, seemed to us quite agreeable enough to stick around for a pint or two.

For the next few miles, little walled-off fields belonging to small farming homesteads could be visible between the stone mills, rising away up the sheep-dotted hillsides. We continued on our way to Hebden Bridge, which at one time gave employment to the folk of the town in eighteen different mills and clothing factories. Today, there is not much evidence left of that life. Instead, the town is a thriving artistic community, and has earned the name 'the lesbian capital of Europe'.

As you can imagine, it is a lively place, with shops full of attractive home-produced goods: shawls, pottery, woodwork, delicious food. There is a little theatre, into which we were invited, and we saw a rehearsal of a play performed by some very talented local children. They talked to us, and we were relieved to hear that, in spite of their all being very good, only one girl had set her heart on going into the business. We wished her luck.

The year after we did the Rochdale programme, Hebden Bridge bore the brunt of the terrible floods that attacked whole

The sights along this stretch of water are as idyllic as anything else Britain has to offer.

areas of the UK, but particularly the Lancashire–Yorkshire border. Streets of houses were demolished, and in some cases quite swept away. The banks of the canal collapsed, roads and railways were closed. Nobody could claim insurance against the flooding: apparently, the area was uninsurable, simply because of its liability to get flooded.

The nice people who ran the theatre asked us for help, so we went up and did a show for them. The shell of the building was still intact, but everything inside had gone and was slowly being replaced, as and when they could afford it. I was immensely impressed by the generosity of the townsfolk in helping out neighbours who had suffered slightly worse than themselves.

We met a lady who still runs a horse-drawn tourist boat on the canal – one of very few in England who still offer that service. Pru made friends with her horse, Bilbo Baggins, and we were reminded that, in commercial-boating days, Bilbo would have been required to pull a pair of barges laden with the equivalent of twenty-five cartloads, and would probably think nothing of it.

After Hebden Bridge, the mills and factories at last gave way to rolling open country. Occasionally, a handsome villa could be seen, with a well-kept lawn. This reminded me of a house up here owned by the proprietor of a carpet mill, who had allowed us to film David Storey's play *The Contractor* on his lawn.

The play entails erecting a marquee for a family wedding: the first act is all about the process of putting the thing up, the second is the party itself, and the third is about taking the tent down again.

It was logical to shoot it in story order: to begin with the lawn would be fresh and green; then the floor would be put over it for the party; and finally, as the tent came down, the floor would be taken away to reveal the dried, yellow-brown grass beneath.

Unfortunately, it came on to rain hard just as we began shooting, and didn't look likely to stop. The only thing to do was to skip the first act, put the floor down, shoot the second act and – it had cleared up by now – go on to do the final part, and reveal the bruised lawn.

But of course we still hadn't done the first act; and this necessitated a team of scenic artists with pots and pots of green paint, recolouring the lawn. The carpet miller was very nice about it.

As we got nearer to the summit of the canal, the locks became more and more frequent. It was tough work for Pru, though at one point she did get some help from a group of village children. A bit further on we were suddenly confronted by a road bridge, from which some netting had been hung in order to block our passage. We tied up the boat and went to have a look. Apparently, a motorist, at speed, had careered into the parapet of the bridge, and this had crumbled and fallen into the water. The bridge was being guarded by two policemen, as clearly it was too hazardous to be driven over; but I didn't quite see why we wouldn't able to steer under it.

No, it was declared a Dangerous Bridge, too dangerous for anyone to pass over or under, sorry. Mike, who in all his years of directing documentaries had come across similar situations before now, suggested that, while any *person* should not be allowed to do so, might not a *boat* be permitted, without its passengers? Sportingly, they agreed; we walked over to the other side of the road, grabbed the bow line, and pulled her through.

A couple of hours later, we got to the top, 183 metres above

sea level. We should feel proud: among all the trans-Pennine water routes, the Rochdale is the only one that manages to do without a summit tunnel. The railway, after gamely running alongside the canal for miles, eventually couldn't keep up with the gradient, and had to give in. The resulting excavation was actually the longest railway tunnel in the world when George Stephenson built it in 1840.

We stood beside a milepost indicating Lancashire one way, Yorkshire the other. Behind us lay Halifax, Huddersfield and Wakefield; before us, Oldham and eventually Manchester.

But that'd wait for another day.

THE LLANGOLLEN

WE ARE HAVING a bit of a problem with Pru's hearing. She has artificial aids, but frequently forgets about them. At breakfast:

The telephone rang, and Pru answered it.

'Hello? Yes. Yes it is. What? Who is this? I'm sorry, this is a very bad line, could you speak up?' She shook her head in bewilderment.

I took the phone. 'Hello, yes. No you don't need to shout, it's all right. Thank you. Tomorrow. Yes, fine. Goodbye.'

'Sorry, who was that?'

'Window cleaner. You haven't got your ears in, have you.'

'Yes I . . .' – feels – 'no, I haven't, I took them out.'

'Why?'

'I don't *know*. Do you want a divorce?'

'No, I just want you to put your ears in.'

'I can hear perfectly well without them!'

'No, you can't.'

'What?'

'Please put them in, then I can talk to you.'

'Yes. Oh, dear. I don't know how you've put up with me all these years.'

'I know, it's amazing.'

'Do you love me?'

'Yes I do, very much. Would you like another piece of toast?'

'Yes please.'

'You can have another piece of toast if you put your ears in.'

'Yes, all right. Where are they? I forget.'

In fact, we usually do find the hearing aids, somewhere, and in any case there's a spare pair, though I haven't seen them lately. But we got off all right to Llangollen, where years ago we had taken our delayed 'honeymoon'. It is a delightful little town, and the canal that gets you there is a waterway of special distinction.

Although we found ourselves subject to the kind of weather for which that part of Wales tends to be famous, our Llangollen episode produced some really stunning cinematography by James and by his colleague Gary Parkhurst, who filmed us from the towpath as we went by, and then went running far ahead with a heavy camera and tripod, in order to catch us appearing again through the trees. In later episodes, that effect was provided by a 'support boat', to travel round us and shoot from behind or ahead of us, or from either side, as we went along.

We joined our Anglo-Welsh narrowboat at Chirk – or, by its Welsh name, Y Waun. There's a tunnel, about 500 yards long, which is shared by the railway running alongside. The shorter Whitehouse Tunnel, a bit further on, has the same arrangement.

From the canal there's a good view of Chirk Castle, completed in 1310 and the domain of the Earl of March, Roger Mortimer. His descendant Edmund Mortimer is borrowed by William Shakespeare for a scene in *Henry IV, Part 1*, featuring the great Welsh chieftain Owain Glyndwr: 'I can call spirits from the vasty deep,' he boasts, to young Harry Hotspur, and gets this reply, 'Why, so can I, or so can any man; / But will they *come* when you do call for them?'

The Pontcysyllte Aqueduct is probably the most famous and most spectacular landmark on the whole canal system, It is 1,007 feet long and rises 126 feet above the River Dee, and the waterway trough is designed with the non-towpath side completely unprotected from about 12 inches above the water level. We have made the journey once or twice before now, but still Pru usually finds this a good time to go below to make a cup of tea.

The construction, by Thomas Telford, is quite revolutionary. The idea of laying an iron trough directly onto a row of stone pillars was completely new in 1795, when the work started, and today, apart from some few minor details such as the renewal of the towpath, the structure stands exactly as it was. The masonry is bonded together with a mixture of lime and oxblood, and the joints in the iron trough are sealed by – can you believe? – Welsh flannel dipped in boiling molasses.

Crossing the famous aqueduct took us a long time. Because of the sun, we needed to be filmed crossing it in a westerly direction; but the narrow structure allows for only one-way traffic, so we had to wait in a queue for all eastbound craft to get through before making our first attempt to cross over into Wales.

There was a party of pedestrians on the slim footpath beside us, and they got into a loud and prolonged conversation with people on the boat ahead of us, thus making our filming impossible. There was a line of boats behind us, so we couldn't reverse and go back; we had to continue right on to Trevor Wharf, turn round, go back across to where we started, and try again.

This time we got about a quarter of the way across when the heavens parted and it bucketed down, the rain driving straight at the camera lens. So that was no good, either. The third time, we finally got it; and the welcome to Wales provided by the pub at Trevor was warm, timely and necessary.

The canal clings spectacularly to the side of the Dee as you approach Llangollen itself; it gradually gets narrower and shallower, so that motor launches have to give way to narrowboats, which in turn give way to horse-drawn boats, and after that you have to walk. At the end of the path you come to the Horseshoe Falls, a great semicircular weir across the River Dee, built by Thomas Telford to provide constant water for the whole of the canal and beyond.

That the canal exists at all is due to the failure of the planned Ellesmere Canal, supposed to link the River Severn with the River Mersey. That never happened, for financial

Opposite: The fabulous Pontcysyllte Aqueduct – world heritage site and bane of our unfortunate cameramen.

reasons; and a northern section from Ellesmere Port was filled in again almost as soon as it was built because of resistance from property holders – thus cutting off access to the main water source. So the Llangollen Canal was cut as a feeder, to get at the water provided by the Falls.

I'm sorry we never got to Plas Newydd, the home of Eleanor Butler and Sarah Ponsonby, two aristocratic spinsters who, when chronicled in 1774, had reached the ages of thirty-nine and twenty-three respectively: 'The Ladies of Llangollen'.

What was so remarkable about them, and why are they so comparatively famous? The phenomenon of female couples sharing a romantic friendship, sharing sensibility, sharing beds, was not uncommon in the eighteenth century. Yet, in their little cottage just outside Llangollen, they entertained Charles Darwin, William Wordsworth, Joshua Wedgwood, Lady Caroline Lamb and the Duke of Wellington. Why? Neither of the ladies was an artist, nor, except for their regular journals, could they be classed as writers. But they had an extraordinary grasp of what was happening in the world, conversed intelligently, had a lovely garden, and were fun to be with.

They remind Pru very much of a pair of ladies who lived in a very small cottage overlooking the sea in Bucks Mills, North Devon, where Pru spent part of her childhood in the war years. The ladies were both watercolourists of some distinction, and, although they kept themselves very much to themselves in their rather Spartan circumstances, Pru says whenever she met them she was much struck by their shared interests, their tenderness and sympathy. 'In

Left: People crowd in to watch a musician at the Llangollen International Musical Eisteddfod. © *Corbis/Getty*

Below: Things stand still for Her Majesty The Queen on a visit to the festival. Note the curtseying women from Brittany, in traditional dress. © *Stringer/Getty*

another life,' Pru sometimes says, 'I think I could have been like that.'

We were not in time to witness the Llangollen International Eisteddfod in July, which was a pity. Pru knew a family, two sisters, who were very much involved with the music festival, and we came up to see them once or twice. They are long gone now, but I do remember them with affection.

Though not so the husband of one of them. He was a local magistrate, I think, not at all interested in music – in fact, the only thing that seemed to stir his excitement at all was to be shooting pheasants. In all weathers he would go out with his twelve-bore and range the hills above Llangollen seeking his prey; but in the close season he became so depressed that one morning he took his gun down from the wall, went out into the garden and shot all his hens.

So the traditional boiled egg for breakfast was no more, and things were never quite the same there afterwards.

Nevertheless, the town still held a considerable treat in store – for me, at any rate. I was promised the chance to ride in the cab of a steam locomotive on the Llangollen Railway, which is a privately owned line forming a rescued part of the GWR route from Ruabon to Barmouth, which had been closed in the sixties. Railways past and present (but particularly past) have always been a passion of mine, and it does seem that the people who strive to restore our canal system often have the same concern for old railway lines. Odd that, in so many cases, the latter was the means that drove the former to extinction.

We walked down to the station. There she was, a

beautifully maintained Great Western 4-6-0. We got aboard as passengers, but on the return trip from Berwyn, where there is no turntable, we had to run tender-first and I, having been invited onto the footplate, stood looking out over the tender into the coal dust and driving rain as we raced along. My childhood passion for steam locomotives has never deserted me.

When I was about seventeen or eighteen, I went on a walking tour with a similarly railway-inclined friend who lived in Merthyr Tydfil, and we went up to Tywyn in Merioneth to discover what was happening to the almost defunct narrow-gauge Talyllyn Railway.

The line was completed in 1864 to carry slate from the Bryneglwys Quarry down to Tywyn, to be transferred onto the GWR. After a while, it also began to carry passengers, and for a time things went well for the railway, though less happily for the quarry. In 1911, both were purchased by Sir Henry Haydn Jones MP, who somehow kept the railway running throughout the thirties' economic depression and the war; but times were hard, and by 1946 the track, locomotives (two) and rolling stock were showing serious signs of dilapidation and the company workforce had been halved.

Even in its penurious days, there seems to have been a happily esoteric attitude among the staff: during the war, the guard carried a gun in his van, not to repel invading forces, but on the lookout for rabbits to supplement his war rations. There were stories of oatmeal being routinely poured into the boiler of one of the engines to plug the

The Talyllyn railway route truly is spectacular.

© Getty

leaks; and, when the brakes of the brake van failed, it was simply uncoupled and left behind. The necessity of stopping the train while the fireman got out to chase sheep off the track continues today.

Nationalisation of the rail network happened in 1948, but didn't include the Talyllyn. A number of independent lines wanted to be included in the Nationalisation Bill, as they saw it as a guarantee of survival; and it is not clear whether Edward Thomas, who had taken over as manager of the Talyllyn, ever made an application and was turned down, or whether he just didn't fancy it.

However, when the full list was published, one of the people who read it with particular interest was the ubiquitous Tom Rolt. Sir Haydn had promised to keep the railway running throughout his lifetime, and it was a commitment he kept until his death in 1950. A skeleton service was still operating, but it was clear that, unless something happened quickly, the end was in sight. Tom Rolt, however, had got together with a couple of ex-railwaymen friends, formed a society of interested parties, and agreed with the executors of Sir Haydn's estate that they should take over the running of the line for a trial period of three years.

People joined the society, they donated, they volunteered to come and work on the many practical tasks urgently needing to be undertaken. In May 1951 there was an official reopening, and from that time the railway has slowly but steadily gone from strength to strength.

In 1952, Tom married Sonia Smith, the canals archaeologist and campaigner, whom he had met years before at a screening of Charles Crichton's film *Painted Boats*. Sharing a determination to preserve valuable relics of the Industrial Revolution, Sonia put her energies into helping Tom to run the Talyllyn.

While over the years there have been many additions and improvements to the line – some new locomotives, a much-enlarged station building at Tywyn Wharf, and even a scenic extension of the track past Abergynolwyn – the original identity of the line has been lovingly preserved, and my favourite old engine *Dolgoch*, built in 1866, has been carefully tended through her advanced years, and kept in running condition. Occasionally, I'm allowed, under supervision, to drive her. Cloud Nine for me.

When the Talyllyn reopened, as a privately run but professionally operated railway line, people regarded it as necessarily a unique phenomenon. A couple of years later, when the nearby Ffestiniog Railway was planning to follow the same strategy, there was quite a strong feeling of dissent. 'We've already got *one* of these independent railways,' people said. 'Having a second would just invite competition, injurious to both.'

Well, it didn't, and there are now well over a hundred such independent 'heritage' railways in this country, and the list is growing.

Tywyn doesn't pretend ever to have been an inviting seaside resort, but a few miles down the road is the attractive little town of Aberdyfi, on the Estuary, and for many years we have been coming to stay there for a few days, and have made friends. It's a place that has withstood the test of time, and one of its little shops, the chemist's, Medical Hall, was the site of a story that made its way

No. 4 engine stops in Snowdonia to take on water, much to the interest of passengers. © *Getty*

into *The Cambrian News* while I was there. I've no idea whether it's true or not, but I love it.

Apparently, one of their journalists came into the shop to buy toothpaste, and was intrigued to find two middle-aged men dressed in walking gear, one of whom had been talking to the other in German, now addressing the girl behind the counter in perfectly good Welsh.

As a journalist, he scented a story. 'Excuse me,' he said, 'I don't wish to be impertinent, but it struck me as very unusual to hear someone of your – pardon me – apparent nationality, speaking our language. Do you know our country well?'

'Oh, yes,' said the one who'd been doing the talking. 'I was here, for a while, in the wartime.'

The journalist put on a sympathetic look. 'Ah. I'm sorry. Were you interned nearby? I hope they treated you well.'

'No, no,' the German assured him brightly, 'I was not interned. I was in U-boat in Cardigan Bay. We are waiting for British destroyer, but it is not coming. We are getting very bored, you know. So I am young man, I am coming ashore in rubber dinghy to get fresh eggs for our U-boat. And I was in love with your country, and I am learning your language – *Cymru am byth!* Wales for ever! – and now I am bringing my friend Albrecht to climb the mountains with me.'

I do want to believe it. West Wales was a long way from getting involved in the war, and I can imagine someone saying, 'Hey, Ieuan, I bet that chap was a Jerry; you should have put another threepence on his eggs, do our bit for the war effort.'

CANAL DU NIVERNAIS

IT WAS CONSIDERED that for the final episode of our first series of *Great Canal Journeys* we should go abroad – a big field of possibilities to chose from. We ought to remain in Europe, we felt: there were plenty of canals in Germany, in Holland obviously, in Belgium and Northern Italy, but the obvious country to start with was France.

A few years before, Pru and I had been on a hotel boat on the Canal du Midi up to Carcassonne, and enjoyed the experience. The other six guests were all American, and we became friends with a nice publisher and his wife from Connecticut. There was also a rather older gentleman from Southern California, and him I found it harder to deal with. At dinner on the first night, he loudly opened the conversation with the words, 'I consider it the inalienable right of every American citizen to carry arms.'

We decided it might be as well to deflect the conversation

to some other topic, at which point he brought out a pack of visiting cards, and passed them round the table.

'I need to solicit support for the Primaries,' he told us urgently, pressing a card into my hand. Good God, I thought, can this man be a senator? A congressman? A surprise presidential candidate, even? It finally transpired that he was actually lobbying for election to the San Diego Water Board, but personally I wouldn't have backed him even for that.

His forty-year-old daughter, with whom he shared a cabin, came ashore with us one day, and we persuaded her to risk joining us for tea in a little canalside café, even though it was full of French people. A young couple next to us were enjoying an animated conversation, and she gazed at them mystified.

'I don't know how they *understand* each other,' she said.

Although scenically the trip had been a glorious experience, we thought we might look somewhere fresh (actually, we did go back to the Midi later, but starting at the other end), and we finally settled on the Canal du Nivernais.

The Nivernais actually predates nearly all of our own canal system: the Duke of Bridgwater came over and advised about its construction in 1784. Its purpose initially was to help the floating of timber from the forests of Morvan National Park, right up to Paris. 'Flotteurs', as they were called, roped together masses of logs, and on top of this huge raft built a little rough cabin to live in while they poled their way, or were carried downstream by the current, as far as Auxerre, thence to be transported north by more conventional means.

The flotteurs were superseded by barges carrying building stone, grain and wine from the region of the Yonne up to Paris, and the canal quickly became an important navigation route and helped enormously with the economic development of the area. That declined with the familiar pattern of competition from the railway.

In the last few decades, now that commercial traffic has had its day, British influence on the growing leisure trade has been considerable. Our boat hired for the programme came from a British-run boatyard; the lady who took us up the spectacular Rochers de Saussois represented a British

Reserved exclusively today for recreational craft, the delightful environs of the Nivernais canal can be entrancing for visitors.
© Rex Features

tourist agency; British hotel and restaurant boats kept passing us on the water. Apparently in the fifties there was an influx of UK tourists into the area, and it was quickly seen that the canal was a significant though unappreciated element of the countryside. It had become moribund after the demise of the working barges, and nobody seemed to have felt the need to provide for its future.

The French attitude to living by river or canal seems essentially different from ours. We like to make our canalside cottage look attractive from the water, with a nicely tended strip of garden if there's room, and perhaps a little boat tied up, or at least a tiny landing stage. Certainly no dustbins. The French perspective is quite different: they want to show their house to the *road*; the canal is round the back.

So a long period of improvement began on the Nivernais, with boat builders, hire firms and restaurateurs from across the Channel giving the nudge to local developers. For us, it was a very pleasant voyage: the boat was large, but easy to handle; the channel wide and well-dredged; and the locks professionally operated, usually by young female university students who, as they saw the boat coming, would lay aside the book they were studying and come slowly forward to open the electrically operated gates. Except between 12 noon and 1 p.m. – that's the inexorable closure for lunch.

Joe and our granddaughter Matilda drove over to join us from their part of France, the Jura, close to the Swiss border. They live in a very attractive old house beside a little river, in a village called Arlay. It is very quiet, and I've often thought it would be a perfect place to which to retire. So

In our best Gallic dress.

it would, but Joe, at the time of writing, is only forty-eight, and a talented writer and translator. His wife, Hedwige, is a very accomplished artist with pastels, and the children are all quite capable musicians – and, I thought, if they lived in Ealing, or Walthamstow, or Motspur Park they could live enormously active lives, doing a lot of things for which they probably wouldn't be getting paid, but would find

89

creative and satisfying. But never mind, the kids are grown up now, and at least Joe will have time to write, in peaceful surroundings. I love them all very much, and want them to be happy.

We went to the typical French local market at Chablis, and bought a great many different mushrooms, also a length of blood sausage, witnessing the manufacture of which was not a spectacle for the faint-hearted. We cooked it on board and had a delicious dinner, and then Joe and Matilda had to leave us.

The next day we visited the vineyards of Mailly-la-Ville, where picking was in progress. The Chenin Blanc grapes for the famous Cremant have by law to be picked by hand, not by machine, so we were happy to help and got quite good at it; and after lunch we were invited to a tasting by the cellarer (again an Englishman), and indeed bought some wine to be shipped home.

Our journey ended at the busy port of Auxerre (some people pronounce the 'x' but I'm not sure that's right: we don't say 'Bruckselles', do we?). We left the harbour, full of yachts, sailing barges and motorboats, and climbed up to the Cathédrale St-Étienne, where long ago Joan of Arc celebrated defeating the English at the Battle of Poitiers.

We said goodbye to our boat, of which we'd grown very fond, packed up, and were driven away to Paris, reflecting on another enjoyable and memorable French experience. And we would soon find ourselves in France again – on the Canal du Midi – but more of that later.

Opposite: The magnificent Cathédrale St-Étienne.
© Rex Features

INTERLUDE

ALL IN THE TIMING

PRU, PARTICULARLY, LOVED being in France. She had been taught French very thoroughly at an early age, and still speaks it fluently (although, she says, in rather an old-fashioned style). She spent quite a long time in France when she was younger, and had a French boyfriend, whom we don't talk about much.

She is fascinated – we both are – by different languages, different dialects, different ways in which people express themselves, and this I believe is very much the key to the success she's enjoyed in television comedy: Miss Mapp in *Mapp and Lucia*, Miss Bates in *Emma*, Dottie Turnbull in the Tesco ads, and of course the immortal Sybil Fawlty. They are all specific, accurate creations, not just popular impressions made to fit.

Her rigour is due, I think, to the fact that neither of us likes 'comedy' to be thought of as a special style of acting.

Opposite: Looking suitably louche in *Brass*, a comedy I worked on in the eighties.

Some casting directors are in the habit of saying about an actor, 'Yes, but does she do comedy?' *Do* comedy? If she's in a play that has a funny line, and she says it, and gets a laugh, then she Does Comedy, that's all there is to it. She has got the laugh only because she was *real*, believable as a character and in the situation. It's the same as any other kind of acting.

Of course I am talking about comedy *drama*, not the shows where people are just there to make jokes in front of an audience; they're really just a form of stand-up, and very funny some of them are. But they're different.

And that brings me to a subject that Pru and I don't actually agree about. She doesn't mind playing in front of studio audiences, and of course *Fawlty Towers* was done like that, and worked wonderfully under John Cleese's stewardship. But, as soon as I hear a studio audience laughing, I actually cease to believe in what's happening. Where *are* those laughers? I want to know. Sometimes there's a scene in a telephone kiosk, and you hear two hundred people laughing. There's also a problem of *pace* – yes, I know *Fawlty Towers* solved it, but, then, John's a genius – in that, if someone says a line and gets a huge laugh, what does the vision mixer do?

Stay on the chap who said the line, and hope he follows it up with making an interesting face, or cut early to the other person who will have to spin out his reaction? Either way, it's slowing things up. Also, by having to find room for an audience you're giving up half your studio space, which could be used to give more depth and credibility to your set.

We did a show called *Brass* for Granada, in Manchester; it was a spoof, sending-up everything that had ever been written, filmed or painted about the North of England. Nothing was sacred: D H Lawrence, J B Priestley, *Love on the Dole*, *Hobson's Choice*, L S Lowry, *Coronation Street*. As a spoof, it had to be authentic, had to be real. It was also extremely funny and must be played very fast.

Clearly it wouldn't work with a studio audience, but that was how Granada initially wanted to do it. It's a *comedy*, they said; if they don't hear people laughing, how will they know it's meant to be funny? Oh, I think they will, we said. Fortunately it worked.

Two of the most successful TV shows of the nineties broke the mould: *The Office* and the glorious *Royle Family*. All played truthfully within the situation, like all good acting; and to my mind wonderfully and consistently funny – *without a laugh track*.

But, of course, *Fawlty Towers* is special. When I was watching one of the repeats it suddenly came to me how important a part the *set design* played in the show: the staircase and the long bedroom corridor take a bit of time to travel, so that even the biggest laugh has quietened in time for Basil to come in with his next line immediately. That's brilliant.

The first time Pru met John, he was in bed with flu, in his rather grand flat in Hyde Park Gardens.

'Do you like the scripts?' he asked.

'I think they're brilliant.'

'Any questions?'

'Why did they get married?'

'Oh, God! I knew you'd ask that.'

They chatted, and Pru got the part. She didn't know how John (and Connie Booth, his co-writer and wife at the time) had originally imagined Sybil 'a bit differently', they later told her. But, by the second day of rehearsal, they were convinced that Pru had made all the right choices, and were very happy.

The idea for the series came, as most people now know, from a hotel called Gleneagles in Torquay, where John and his colleagues were staying during the filming of *Monty Python*. The hotel proprietor, Commander John Sinclair, was so offensive and negligent that the rest of the cast moved out quickly; but John and Connie, sensing material for a sitcom, bravely stuck it out.

Eric Idle, before he left, mislaid his briefcase. After a while it was discovered abandoned on the other side of the garden wall. Sinclair had put it there because it contained an alarm clock, which ticked faintly, and Sinclair declared it to be a bomb.

Quite a number of incidents that occurred during their stay found their way into *Fawlty Towers*. When John made the surprising announcement that the series would end after only twelve episodes, he proclaimed that he was now emancipated from all the hatred he felt about British hotels.

I used to describe Pru as a workaholic. An idealistically motivated alcoholic. To work, or have the immediate prospect of work, was to her as essential as food or drink. I'm completely different. Work to me is not the be-all and end-all of my existence: rather it is the related benefits of an actor's life that attract me more than actual performance

– the people we meet, places we go to, things we discover, scraps of information that generally remain in the head long after the actual lines are forgotten. We learn, rudimentarily at least, to do practical things we wouldn't otherwise be asked to do: ride a horse, conduct a symphony orchestra, drive a steam engine, print hand-block wallpaper, pass oneself off as a transvestite. It has always seemed to me that an actor at a party should be capable of carrying on a conversation on almost any topic, for about thirty seconds. After that, we have to excuse ourselves and go and get a drink.

A few years ago, I received a rather unusual enquiry that added to my store of unexpected challenges. It was from the Vatican. Now I am not a Roman Catholic, and being told that my services were required to commemorate the thousandth anniversary of the Martyrdom of St Cyril and St Methodias didn't give me a lot to go on. What did they actually want?

Well, there was to be an International Cultural Symposium to be held in the Vatican, and among the many offerings from different parts of the Catholic world His Holiness had asked for a British actor to read Thomas Beckett's Christmas sermon from T S Eliot's *Murder in the Cathedral*. Obviously, they must have searched through all our Roman Catholic performers, starting with Sir Alec Guinness, and found none of them were free, so somehow, by scraping the Protestant barrel, they had come up with me.

I pictured myself standing in evening dress at an elegant Bernini lectern in an intimate little salon adorned with exquisite frescoes. Sitting before me attentively would be the Pontiff in immaculate white, flanked by the crimson

Opposite: Pru's working life has seen her take on every kind of role with absolute aplomb, but she will, inevitably, find immortality as Sybil Fawlty.

outlines of perhaps a dozen of his best-loved cardinals, faintly discernible in the golden candlelight.

No. The modern public auditorium in the Vatican holds 8,000 (seated) or 15,000 if two-thirds of them are prepared to stand. The building is so long that the far ends are shrouded in mist. The actual stage, across which I am required to process in am-dram archbishop's costume, preceded by four altar boys and swinging a thurible, is gigantic.

I wasn't having much luck with the thurible. In order to puff the incense properly, I had to hold onto its chains with two fingers while using a third to agitate the bellows. My altar-boys had been introduced to me in rehearsal as 'il Archivisco' and had knelt down to kiss my ring, but, when they found I didn't have one, they lost interest and went back to their comics. Now, though, watching my pitiful efforts as an incense swinger, they fell about with laughter, and it was suggested that I take the thing home with me to practise.

As I paused at the City gate, one of the Swiss Guards noticed a thin trail of smoke issuing from a supermarket bag on my lap. He asked me to get out and come into the guardroom. Inside the bag he found this beautiful example of episcopal hardware. It wasn't real gold, of course, studded with real rubies, but he wasn't to know that. Explanation was impossible: he knew no English, and my Italian is limited to a few restaurant menus and a bit of grand opera; but then I thought, Swiss Guard, maybe he speaks French.

So I tried, '*Moi, je ne suis pas voleur; je ne suis qu'un acteur.*'

His face lit up. '*Ah! Comedien anglais, vous?*' he asked. '*Connaissez-vous Benny Hill?*'

'*Mais oui!*' I lied. '*C'est mon grand camarade!*' After that we were all right. Although I never met him, or even watched any of his shows, I have since found that Benny Hill is an effective name to drop in timely fashion if you're in trouble, almost anywhere in Europe. Nowadays, Mr Bean will produce the same result.

PART 3

FORTH & CLYDE
AND UNION

PART 1: AULD REEKIE AND THE
ART OF FINDING A DRINK

PROFESSIONAL WORK was proving very difficult now for Pru. Although her memory for events in her early life was as strong as ever, she had difficulty in recording conversations that had happened even a few minutes ago, unless reminded. We still managed to perform poetry recitals together, and she would have been a useful performer on radio, if producers had felt they had the patience necessary to steer her through a day's broadcasting, accompanying her to the loo, the restaurant and her taxi home. She felt occasional fits of fury about her memory problem, but, as it was a memory problem, she soon forgot it.

Meanwhile, I had just committed a year to *EastEnders*. Some people have a rather snobbish attitude towards performing in TV soaps. I don't know why. Acting in them seems to me to be following exactly the same principles as acting in Shakespeare, or Chekhov, or Noël Coward. You have to think quite *quickly*, of course, because there's a lot to get through on any particular day, and you need to come with a very clear idea about your attitude to a scene; there's no time to second-guess. But I like that.

The other thing I like about the popular soaps is that they are watched by people all over the country, and deal with issues with which they are familiar. The locale may pretend to be the East End, or Manchester, or the Yorkshire Dales, but really the stories, and the situations that give rise to those stories, are common to everyone. The universality takes the emphasis off London, where, at least in our industry, there's a growing feeling that, unless you live, work and sell your wares within the metropolitan area, you haven't really arrived.

We were well looked after in *EastEnders,* the scripts were very good on the whole, and the writers responded to what you might be trying to do with your character: change them a bit, open them up to different, unexpected influences.

The only thing that I found slightly disappointing was the off-duty life. When you're in a play somewhere, or away on location, the ability to meet your colleagues over a meal or a drink or an outing somewhere is surely important and useful. If the only chance you have for converse is a brief chat between takes, that's not the same thing.

Most of the Albert Square regulars have settled for buying nice houses within easy reach of the studios at Elstree; they have families, small children, and, when they've 'wrapped' for the day, all they want to do is to get into their cars and drive off home. I had to be driven right across London to Wandsworth, and would probably get home far later than

General photography usually helps set scenes and evoke locations – in this case, a stunning Scottish waterway.

they; but I would have welcomed the chance of a quick pint with the others before I left, just to share a few thoughts about the day's work.

The BBC very kindly managed to schedule things so that in August 2014 I could take two weeks off to go off to do the next two Great Canal Journeys.

People are often surprised to hear that we shoot the episodes in a week: six days with Pru and me, and the last day reserved for 'GVs' [general views] – scenic photography, usually involving aerial shots and mostly using drones. If our boat appears in any of these shots then of course two stand-ins have to be procured, to look like Pru and me.

Sometimes this is very difficult: my bald head is a problem unless I'm wearing a hat: filmed from above, all you see is the hat. We lend them the clothes that we were wearing, which are laundered and posted back later.

Of course the production week forms only a small part of the labour that goes into making the programme. We have already talked about the extensive preplanning, but, after we come back from the shoot, Mike has to immerse himself in weeks of editing. Then we have to decide on the off-camera commentary that Pru and I are to speak, and that needs to be tailored to picture, and then recorded by us both. A final editing polish, and the programme is ready to go on air.

Good. So where are we to go next? Having already paid a visit to Wales, we thought maybe it might now be the turn of Scotland. We rather liked the idea of starting in her capital city.

We were no strangers to Edinburgh. Starting in, I believe, 1966, I have been involved in more Edinburgh Festivals than I'd care to remember. In 1969, we reopened the Assembly Hall of the Church of Scotland with memorable productions of *Richard II* and Christopher Marlowe's *Edward II*, followed three years later by my first attempt at *King Lear*. The last thing I did in Edinburgh was a Fringe play about bomb disposal, in 2005.

Of all the productions I must have seen, all the music I must have heard, all the people I must have met during the Festival, one little memory still stands out. On the last Sunday morning after the Festival has closed, the 11 o'clock London train tends to be packed, and one year the railway generously decided to provide a relief train, to leave shortly after the other one.

I was one of the first people to hear about this, and I found the train, quickly boarded it and sat in the dining car. There was one other occupant, at a table some distance away: it was the conductor Carlo Maria Giulini. I trembled. For me already a figure of heroic proportion, last night I had been at the Usher Hall for his shattering rendition of the Verdi *Requiem*.

He smiled across at me, and I tried to smile back. I desperately wanted to go over to him, perhaps kneel at his feet, and tell him of my unbounded admiration over the years, but seeing him now, close to, I was tongue-tied with embarrassment, and buried myself in a book.

The waiter came, and took his order for lunch. When it arrived, he ate a little of it, and laid down his knife and fork. The hovering waiter, obviously a devotee, asked nervously if anything was wrong. A murmured request led the waiter to summon the entire kitchen staff (three), who stood round the table in deferential silence while Giulini gently explained how vegetables were normally cooked in his native Italy. They bowed respectfully, and took away his plate to try again. They did, and it was better, and the maestro gravely congratulated them.

The essential cosmopolitanism of the Festival was severely tested in the 1960s when on Sunday you couldn't get a drink in Scotland unless you were 'a bona fide traveller' staying in a hotel. This was a painful situation for a lot of participants arriving on Sunday morning: foreign theatre companies, groups of students on the Fringe, whole

In full regalia for a Festival play – devoted acolytes in tow. 1970.

international symphony orchestras – it must have seemed extraordinary to them to be met with such studied lack of welcome.

One year, I had just arrived with a touring production that was to open next day at the Lyceum. A number of us congregated for coffee at the Festival Club in George Street, and after a while our distinguished leading actor – I'll call him Robert – suggested we go upstairs for lunch.

Robert was known as a man who valued a decent bottle of Burgundy, and I thought it worth reminding him of the Sunday prohibition. 'Oh, never mind that,' he said loftily, 'leave it to me. Lunch is on me.'

There were six of us, and the generous Robert selected a table and asked to be brought the menu. 'Now, we've got a lot to discuss,' he told us, 'and we don't want waiters coming and interrupting us every few minutes; so what I suggest is we all decide on everything we want – soup or whatever, the main meal, and dessert – and order it all together.' So we did that, all six of us, everything; and the waiter wrote it all down and went away.

Before any part of the meal could arrive, Robert grabbed the attention of another waiter, and asked, 'Could I see the wine list?'

There was a pause. 'The wine list?'

'Yes, the wine list. Could I see it?'

He shook his head. 'I'm sorry, sir. It's the sabbath.'

'Yes, I'm aware it's the Sabbath. But I'd like a drink, you see.'

What was going to happen? Robert could be irascible. The five of us looked at our toes.

'No, sir, I'm afraid that's not possible.'

'Isn't it? Oh I see. *Right.* Up we get, gentlemen . . .'

Avoiding the waiter's eye, we got to our feet and followed Robert across the room and down the stairs of the club. As we passed the kitchen we could see a stack of dishes being loaded onto a trolley – our lunch.

Our leader marched us across the street and into the George Hotel, where he was staying, and where he had already reserved a table for all of us. We sat down, and our splendid repast was washed down with a very respectable Gevrey-Chambertin.

'That'll show 'em,' he growled.

FORTH & CLYDE
AND UNION

PART 2: THE OTHER WOMAN
IN MY LIFE

BEFORE SHE MET ME, Pru had been very much in love with a young man called Bill Blackwood, who lived at Gogar Mount in Edinburgh and edited *Blackwood's Magazine*, of whose founder, William Blackwood, he was a descendant.

The house was a Regency hunting lodge, with a large circular central hall onto which four different Victorian annexes, to the north, south, east and west, had been added. These four areas were separately occupied by Bill's mother, her nurse, the housekeeper and Bill himself. They all disliked each other, and never met. In the rather lovely Regency sitting room in the centre of the building, behind an ink-stained sheet hanging from the ceiling, Bill had set up the press to print the magazine. Into the other rooms he seldom went, and, as the housekeeper therefore saw no reason to clean them, they gradually became unusable and

Bill would simply close them, for ever, and put the key back in the kitchen with all the others.

Pru and I used to visit him whenever we came to Scotland but, in 1994, Bill was driving back from a Hunt Ball when he came down a hill too fast, failed to turn, hit a brick wall with some force, and was killed outright.

After his death, Gogar Mount happened to pass to Mrs MacKinnon, the heir to the hundred-year-old Drambuie fortune. Apparently, the secret recipe for the liqueur was kept in a locket round her neck, and every week she went out to a little shed in the garden to mix the concentrate. When she died, at last the house was sold, and Drambuie was bought by William Grant and Sons in 2014.

During the winters, the conditions can become rather difficult in these parts, with huge corridors of frost and snow. © PA Images

The canal we were now proposing to explore is in fact a combination of two canals. Thirty-two years after the Forth and Clyde was built to cross the country from Glasgow to Grangemouth, a route was planned to connect its eastern end to central Edinburgh by means of what became the 'Edinburgh and Glasgow Union Canal'. We went into the city to try to find it.

Turning off at Fountainbridge, right in the busiest part of the Old Town, and avoiding the buses, suddenly you find yourself beside some water. It's the terminus of a canal. That's not an unusual phenomenon: in many places a canal will have been purposely brought to reach right to the heart of a town, where the loading and unloading of goods could happen most easily, and the canal wharf would have become the city's focal point. Now that the cranes, winches, warehouses and grain stores are no more and have given way to supermarkets, hairdressers' and coffee shops, the canal looks oddly out of place.

Somebody, we saw, had had the shrewd idea of buying a narrowboat to accommodate Festival patrons footsore from toiling up the Royal Mile and exhausted by invitations to visit a hundred Fringe events, and take them a few miles down the Union Canal, passing the back gardens of the Edinburgh outskirts, and up to the green spaces of Ratho.

We did this trip, happily, and then transferred, at a charming pub, to our hired boat for the week. Unfortunately, the boat had been moored facing Edinburgh. To turn an unfamiliar boat with about a foot to spare while being watched by a full complement of patrons outside a busy pub was not a joyful experience, but we managed it.

The sign on the side of our craft advertised its origin, from Anglo Welsh Boats. It was thought by some that in Scotland the sight of this logo might confuse the viewers. So the name was painted over, and this, we heard, made Anglo Welsh very cross.

The landlord at the inn at Ratho had very kindly supplied us with all kinds of local produce – bread, pâté, eggs, honey, tomatoes; we had a delicious dinner and settled down for the night as soon as the geese finished their conversation. Next morning, we came to Linlithgow Canal Basin, surrounded attractively by old canal workers' houses, stables and bridges, and also by the Canal Centre Museum and offices.

Here we were able to attend a meeting of the society looking after the canal, and were shown a lovingly preserved fifteen-foot-long map: the engineer's original survey of the route, right up to the point where the Union originally turned onto the Forth and Clyde through eleven locks with a total rise of 115 feet, needing the passage of 3,500 tons of water each time the locks were used.

Not surprisingly, by the 1930s not many vessels were of a mind to do this, and before long all eleven locks were dismantled. This meant effectively that Edinburgh was cut off from Glasgow as far as traffic by water was concerned, and the Forth and Clyde closed completely at the end of 1962. As a result, the Union Canal also fell into disuse along much of its track, and, in order to keep the water flowing, culverts were used in several places: at one point the canal became just a pipe under a housing estate.

In 1997, the British Waterways Board decided that

something had to be done. A return to coast-to-coast navigation was deemed impossible, but the Millennium was looming, and the sale of lottery tickets had already achieved some astonishing success in helping 'worthwhile projects'. So the two canals were being gradually reopened, and public attention was focused simply on how to link them.

The Antonine Wheel, as it was originally called, was designed to raise boats 79 feet from one canal to the other by means of enclosing them in 'caissons' (the Anderton Lift on the Trent and Mersey already does this) and somehow getting them up to the higher level. The final design of the Falkirk Wheel is astounding, I think: so graceful, reminding people of a bird, a Celtic axe or a ship's propeller, whatever takes your fancy. The size and weight of your boat doesn't

The astonishing achievement of engineering and design that is the Antonine Wheel – now known as the Falkirk wheel.

matter: about the only thing I remember of school physics is Archimedes's Principle of Displacement: the water weighs the same as the boat, 500 tons to a caisson. The arms of the amazing machine balance each other, and rotate by an electrical charge equalling what's necessary to boil just eight kettles.

Just before we got to the Wheel there was a tunnel that Pru, no admirer of such excavations, declared the most unpleasant she had been through – ever. Her opinion is based on having stepped up on deck just as a massive deluge of water cascaded on both of us. As if to make up for this behaviour, the tunnel offers an interesting geological anomaly. Stalactites, as we know, normally take thousands of years to form; but the ones here in the tunnel are already substantial, though they date, of course, only from when the canal was built two hundred years ago.

A little way from the Wheel you see another extraordinary example of Scottish imagination: the Kelpies, two gigantic horse heads sculpted of fragmented steel. They are very beautiful and proud, and I was pleased to learn they were modelled on two old shire-horses that had previously worked on the canal.

We followed the Forth and Clyde as far as Stockingfield Junction, where you can push on to Bowling, on the actual Clyde, or turn left and go right into Glasgow. That's what we did, and what did we see? We saw the Other Woman in my life.

I'd better explain. The Other Woman is a 240-foot paddle steamer called *Waverley*, built in 1947 for service on the Clyde and destined for the scrapyard, had it not been for

the intervention of an organisation appropriately named the Paddle Steamer Preservation Society, whose chairman, Douglas McGowan, agreed to purchase the vessel for £1 in 1974. I am an active member of the society, and I have been deeply in love with our beautiful purchase ever since. Sorry, Pru. (No, actually, Pru is quite fond of her too.)

We went aboard, and set off 'doon the watter'. After a little way, we passed an inlet on the starboard side: the canal entrance at Bowling, now full of small sailing vessels eager to get out and catch the breeze, but nervous of *Waverley*.

My devotion to paddle steamers was inherited from my

Pru and I have a terrible habit of getting in the way of the views, I'm afraid.

father. I don't know where he got it from, but after he died I went through accounts of some of the theatres he had played in, in various places, before the war. And a curious pattern began to emerge. Bristol, Liverpool, Eastbourne, Weymouth, Margate, Southampton, Ilfracombe. All places where fleets of paddle steamers were based, or to which they frequently sailed during the season. In early life he was a popular young actor, and most provincial managements would have been glad to have him in their company; so did he just choose places where he could watch the steamers coming in, stand on the jetty and hail the captain on the bridge (even in later life, to Dad such men were as gods), and on Sunday sail off with them down the Solent, or along the Dorset Coast, or over to Lundy Island, going down to watch the pulsating cranks in the engine room, or just listening on deck to the rhythmic churn of the paddles?

I think he did, and years later, when we were still living in Bristol after the war, he took me down to Hotwells on an April afternoon in 1946, to await the reappearance of Messrs P & A Campbell's paddle steamer *Ravenswood*, returned after her war service as a minesweeper, and once more resplendent with white funnel, blue house-flag and polished brasses, to take us on the first passenger excursion a steamer had made from that city in seven years. As I saw her approach the landing stage, the grey filter of austerity that had shrouded half of my young life was lifted. Now the war was *really* over.

The White Funnel Fleet enjoyed two more decades of optimism, as four more of their war-torn heroes were brought back to commercial life in the Bristol Channel, together with two proud, newly built vessels, the *Bristol Queen* and *Cardiff Queen*. While we remained in Bristol, the family used to take a fortnight's annual holiday in Uphill, outside Weston-super-Mare, and I used to spend most of that time on the Campbells' steamers, crossing over from Weston to Cardiff and Barry, up the channel to Clevedon and Bristol, or down the coast to Lynmouth and Ilfracombe.

We disembarked *Waverley* at Greenock. It's a shame we didn't have time for the full cruise to Rothesay and up to Tighnabruaich, but it was good to see that most of the passengers were going the whole way.

The day excursion by sea was something that strongly

The magnificent *Waverley* – the other woman in my life – at the pier.

Opposite: Our Ruby Wedding
anniversary aboard
Waverley.

appealed to people in the fifties and sixties; but, as soon as car ownership became general, their concern was much more for the destination than the voyage. One by one the ships were laid off, sold or scrapped; and this didn't just apply to the Bristol Channel: Southampton's Red Funnel Steamers were finding that their clients preferred the Southern Railway car ferry to Cowes to their own regular cruise to Sandown, Shanklin, Ventnor and round the Isle of Wight. Cosens & Co. of Weymouth, the Isle of Man Steam Packet Company in Liverpool, even the Eagle Steamers in London, were suffering the same fate.

When, years later, the *Waverley* came to be rescued, she was now the last seagoing paddle steamer in the world. Her new owners were aware that to drum up support for this singular piece of history, it was not enough to rely on public nostalgia. Nostalgia works only if it brings back actual memories; but the people who can actually remember paddle steamers are necessarily few, and clearly advanced in years. They probably won't be around much longer. Fortunately these days a new clientele is gradually emerging: young families, married professional couples and foreign tourists, as well as elderly anoraks like me. To maximise revenue, *Waverley* is no longer confined to the Clyde: she operates all round the country with seasons in the Thames Estuary, East Anglia, on the South Coast, the Bristol Channel and in North Wales crossing to the Isle of Man.

Pru and I chartered her in 2003 for a big party to celebrate our Ruby Wedding anniversary, and we did the same thing ten years later, for our Golden. After that, who knows?

THE OXFORD
CANAL

WE RETURNED TO OXFORD to begin the next episode of *GCJ*, and took Mike and the others to visit the Eastgate Hotel, where a significant event in the courtship of Prunella Scales and Timothy West had taken place long ago. Then we turned to our beloved canal, looking very rundown and graffiti-sprayed at this end. After Hythe Bridge, a plywood wall had been erected all along the right-hand bank, concealing the sight of Worcester College and the cricket ground.

Philip Pullman, author of *His Dark Materials* and an authority on canal life, came on board and told us that he had become part of a group to recreate the original canalside landscape of Jericho, where in earlier days we used to tie up our boat to be serviced and refreshed by College Cruisers' boatyard services. The company's proprietors were at the moment reduced to living in two shipping containers, and the Byzantine Church of St Barnabas was currently screened off while negotiations were going on between developers, Jericho-dwellers and the local authorities; but Mr Pullman, a man with whom I think it would be very foolish to argue, assured us that it would all be sorted out.

Urban Oxford suddenly gives way to the countryside as you pass under the A34 and A40 bridges in close succession. We stopped off at Thrupp, a genuine canalside village, and were joined by Sam, who would be with us for the day.

Thrupp Lock was where our other son, Joe – aged about seven, I suppose – fell in. Now, everyone falls into the canal at some point; it's a natural thing to do, and the channel is usually only about three feet deep. But in a lock, of course, it's different.

However, Joe had already shown himself to be a good swimmer, so we weren't too bothered. He came up again, and then abruptly sank. Why? I was just preparing to go in when he reappeared, and we got him aboard. Then we heard the explanation. '*Never* let go of the windlass,' he'd been told. 'If you lose it, we're stuck.' When he fell in, Joe had two of the heavy things, one in each hand, and they sank him. Dutifully, he wouldn't let go, but finally, as the lesser of two evil consequences, he decided to drop them, and surfaced.

We were quite grateful, and not at all cross, because we'd met a man who had one of those underwater powerful magnets, and by fishing around for a while he finally came up with both our sunken windlasses and a curious assortment of other things.

In Kenneth Grahame's *The Wind in the Willows*, Ratty

Joe always did like to dive into our canal endeavours headfirst, so to speak.

is probably imagined on a quiet reach of the River Thames when he introduces Mole to the delights of 'messing about in boats'. But Mr Toad, forever on the lookout for new recreation, comes across a working bargewoman in the story, so I think Grahame must have been equally acquainted with our canal.

He went to school at nearby St Edward's, and, while Pru went to look at an adjacent nature reserve, I was shown round by the headmaster, or 'warden' as he is styled, and saw the large mural that Grahame had painted on the restaurant wall, depicting the world inhabited by the river folk of his story.

Lower Heyford was the place where Lynn Farleigh kept the boat that she lent to us for that momentous holiday in

Meeting with Stephen Jones, head teacher of the school that *Wind in the Willows* author Kenneth Grahame attended. With all the time in the world to loll around on riverbanks, Ratty and Mole really were living the good life!

1976. I can still remember that morning, the four of us getting off the train having changed at Oxford, laden with bags of quite unnecessary goodies, and crossing the railway tracks to get a sight of our new holiday home. Lower Heyford is still a busy and popular boatyard and mooring place, with the whole of the canal ahead of you, but easy to get to the Thames from if that's your inclination.

A few miles on we were met by Tom Rolt's widow, Sonia, and we got together an elaborate tea for her on the bank.

We pressed her to talk about her wartime experiences among her fellow women volunteers who kept the canal boats running with their essential cargoes of coal, iron ore and machine parts while the men were away at the front. The girls were jokily termed 'Idle Women' by those who remained. Idle? Anything but. She had extraordinary energy all her life, and I remember coming to see her in her fourteenth-century house at Stanley Pontlarge. (Who *was* Stanley Pontlarge? Surely an overweight, gay, Manchester costume designer.)

Sonia was then in her late eighties, and we were putting together an illustrated talk for the coming Bath Literary Festival. I'd brought some lunch; we worked, we ate, we had some champagne, we worked some more with Sonia coming and going and fetching a lot of books and photographs; and finally, at about half past five, I said, 'Well, that's enough, I think, don't you? It's been a long day, and you'll want to put your feet up.'

'Oh, no!' she said in surprise. 'Gracious, no! I'm going to the theatre.'

'*Are* you?' I said. 'To see what?'

'*King Lear*. In Stratford.'

Shortly after we filmed our canalside chat, Sonia sadly died. She was 95. Veronica Horwell, in her *Guardian* obituary, called her 'the Grande Dame of Britain's Waterways'.

A bit further on, it was good to meet Jo Bell, the recently created Poet Laureate of the Canals. That's a great idea: the special images and sensations of canal travel do offer themselves to poetry.

Shortly after hearing her clever little piece about a kingfisher, we actually saw one right in front of us, swooping down for a fish. You can go whole weeks on a canal without seeing a kingfisher, and now, suddenly, there it was, in a lustrous flash of colour. James caught it on film, brilliantly, in mid-flight.

The Great Western at Aynho Wharf is a pleasant pub, full of railway memorabilia, and I got to thinking about the proposed electrification of the railway, which, as I grew up in Bristol, became my alma mater. Of course, the Great Western has lagged for years behind all the other national routes that were electrified years ago, and, when it finally happens, no doubt there will be all sorts of benefits.

And yet. The tracks to the West of England lie pleasantly hidden from view among the hills and dales of the Mendips and Cotswolds. Now the landscape will be regularly marked out by steel masts bearing the overhead lines.

But wait a minute. As I write this, the citizens of Bath have forbidden such a thing to happen in their treasured Georgian precinct. Electrification of the main line from Paddington must stop short of Bath Spa! A stimulating concept . . . But no: they'd simply run out of money.

Tooley's, where we tied up in Banbury, is the oldest continuous working dry dock in Britain. Our friend Matt, who runs the place now, showed us round the restored 1930s workshops and the 200-year-old forge, where a blacksmith was fashioning metal ornaments, door furnishings and various household appliances. Pru and I were allowed to have a go, with rather limited success. I made something that ended in a sort of hook.

At one time, long wooden canal boats were built at Tooley's, but, when steel took over as the preferred material, their work became confined to repairs. It was lying beside this boatyard that Tom Rolt found an abandoned old horse-drawn barge, *Cressy*, and decided to install an engine, fit her out and live on board.

This was achieved under the watchful eye of old Mr Tooley, and, when at last the work was finished and *Cressy* cast off for the first time, he stood beside Tom on the aft deck, in his Sunday suit and best bowler, to 'give him a hand' as far as Cropredy. And so, *Narrow Boat* was written, published in 1944, and the rest is canal history.

We moored along a line of boats on the pathway in front of the Castle Quay Shopping Centre, and we found our way through it in order to discover the actual town. I'd been asked to appear, as a joke, at the National Town Criers Championship to be held in the market place, and was duly fitted out with a robe, tricorn hat and chain.

There were about twenty proper town criers, from all over the country, and they each had to stand on a dais, ring their bell and read from a scroll. Sitting behind a table,

It was especially lovely to revisit the canals of Oxford – after all, we had our first ever date in this part of the country.

131

with pen and paper, were three serious-looking men in suits: the judges.

I was petrified. Who had got me into this? The town criers were all giving what you might call 'character' performances – some had thick beards, others long twirling moustaches – and they yelled it out at the tops of their voices, getting enthusiastic applause.

Then it was my turn. I hadn't been given a script, my scroll was blank. So I thought I'd shout the praises of the Oxford Canal.

'Oyez, oyez, oyez!' I said it was a pity the townsfolk couldn't actually *see* the canal, because of the shopping mall and monstrous car park, but it was there all the same, and I urged people to walk through and have a look sometime.

I got applause from some quarters, but black looks from the judges, who I think were probably members of the Banbury District Council. Then it was over, and Pru and I thankfully went to have a drink in a marvellous old pub we knew, the Reindeer, in Parson Street.

It was good to return to the old Oxford, and I'm sorry that, for reasons of time, our journey had to be cut short at Banbury. The canal winds on attractively north along the Cherwell Valley to the village of Cropredy, which boasts two very different historical events.

The annual Folk Festival started in 1979 when Fairport Convention held their last concert here by the canal. Three centuries earlier, on 29 June 1644, Oliver Cromwell's forces under General Waller attacked Cropredy Bridge in an attempt to force a way through to Oxford. They failed, despite their superior numbers, and Waller's soldiers and artillery were captured by the Royalists.

The Battle of Cropredy Bridge is still fought by enthusiasts every 29 June, in silver-sprayed woollen chain-mail, armed with whatever weapons can be locally fashioned, and mounted on what equestrian assistance is available. We've watched the battle sometimes, and it's pretty good, unless it's raining.

Rain, of course, being the occupational hazard of British waterways. Well, there has to be at least one drawback.

CANAL DU MIDI

THE NIVERNAIS EPISODE of *Great Canal Journeys* had apparently been very popular, so we thought we ought to go back to France and perhaps, under different circumstances, revisit the Canal du Midi, on which Pru and I had travelled in the hotel boat some years back.

The Midi, which was declared a UNESCO World Heritage site in 1996, is generally accepted as being the *sine qua non* of the French canal system. Providing a route from the Atlantic at Bordeaux to Sète on the Mediterranean, it traverses the regions of Languedoc and Rousillon, and earns its name 'Midi' from seeming to provide the warm sunlight of midday. Which, as far as we were concerned, it did.

We flew from Heathrow to Montpellier, slightly to the east of the Mediterranean end of the Canal, but during the flight there was an announcement that astonishing rainstorms had flooded the city of Montpellier, and, while it was still possible to land at the airport, passengers would almost certainly have to wait there for some time before roads were clear enough to take them to their destinations in town.

There was a groan from the passengers, and the captain offered an alternative: we'll fly on to Toulouse instead. Again groans. The democratic skipper suggested we put it to the vote, and, to our relief, the referendum found in favour of Toulouse, and we spent the night there. We subsequently found that passengers from other flights who had landed at Montpellier had spent the night in a waterlogged athletics stadium.

Early next morning we drove to the little town of Argens-Minervois, where our boat was moored. It was a *péniche*, the standard hire boat for the French canals – roomy, comfortable, well-equipped. I had driven *péniches* before, but every time I seemed to experience the same initial difficulty in steering in a straight line.

France has a long history of working canals, and in that time one name stands out above all others: Pierre-Paul Riquet. Riquet is memorable not only as the creator of the Midi, but actually as the inspiration behind our own canal system; he was publicly known as the Godfather of all Canals. In 1667 Riquet was commissioned by Louis XIV to build a waterway '*entre deux mers*' to avoid encounters with the coastal Spanish and Portuguese; and this was the result.

Nothing was known about hydraulics in the seventeenth century, and Riquet had no scientific training. He was a good mathematician who simply followed his instincts and learned on the job. While the canal was being planned and excavated, Riquet took residence at the Château de Paraza, which belonged to a friend and patron. Pru and I were invited by the present owner to look around. We stood at Riquet's bedroom window and gazed out upon the canal, as he must have done while the work was going on, three hundred and fifty years ago. He was fifty-nine at the time.

Originally entitled the Canal Royale, of course all that changed after the revolution, and that's when it became the Midi. Our host showed us some original surveyors' plans of

The views from Riquet's old bedroom really weren't all that bad, actually.

the canal, and generously plied us with some very, very old brandy, which Pru tackled gamely but in the end admitted defeat – '666', it was called, and I think I know why. Further down the canal we were given a delicate dinner of lamb's brain fritters, but Pru wasn't awfully keen on that, either.

Next morning, fortified with coffee and a croissant to erase the memories of last night's individual cuisine, Pru began to tell me something of her attachment to a young man years ago who was studying Japanese demonology, and wanted to take her out to Tokyo with him. She was

Putting on the bravest faces we could muster after some brandy – kindly administered by our host at the former home of Riquet – which was older than God himself, and twice as strong.

very much in love with him, she says. Somehow, though, the call of the homeland (she's very much a Home Counties girl, really) swung the balance, and she ended up with me instead. We don't talk about our love affairs, either of us, and I think that's probably a good thing.

In Capestang we were learning something about how, long before commercial traffic began its decline, the canal began a flourishing trade in tourist transport. In 1831, 28,000 passengers sailed from here in a series of boats going up and down the canal, and were put up in a variety of small waterside hotels in sometimes very primitive conditions.

We reached the attractive city of Béziers, the birthplace of Pierre-Paul Riquet, and there, in the market square, is a splendid statue of him. There is a handsome theatre here, too, and up the winding streets to the top of the hill you find the beautiful cathedral perched upon a rock over the gorge.

When we were in Carcassonne years ago we learned a little about the Cathars, that strange sect of Christian Gnostics, originally from the Byzantine Empire, who came over to southern Italy and southern France in the twelfth century, although there are signs of them in the Languedoc a century earlier.

It's difficult to know what influence the Cathars actually wielded in the area. To an extent, all sections of society – nobility and peasants, artisans and merchants – seem to have supported them, and they contributed skills in weaving and spinning to the regional economy. Neither the regular clergy nor their secular brethren appear to have had any serious quarrel with them.

But the bishops, who had been instructed about the need to unify the whole of Christendom if we were to dislodge the infidel from our midst, considered them dangerous heretics. Papal emissaries, headed by Bernard of Clairvaux, were sent out to try to deal with the Cathar leaders, but mutual understanding seemed impossible, and in the end, after armed conflict had broken out, Pope Innocent III decided there could be no alternative to open warfare, and launched the Albigensian Crusade, which took the lives of thousands of Cathars sheltering here in Béziers, including those seeking sanctuary in the Cathedral. In Carcassonne, further up the canal, there was a massacre on a similar scale conducted by English Catholic forces under the Earl of Leicester, Simon de Montfort.

This virtually signalled the end of Cathar influence. There is some evidence of perhaps shared convictions between the Cathars and the Knights Templar, but the Catholics destroyed all manuscripts after the slaughter, every chronicle or scrap of writing that could explain to us the actual teachings that guided their dualist Christian beliefs.

We shook off the charged atmosphere of twelfth-century doctrinal conflict and went down the hill to eat an ice cream and walk around the lovely old town. Back on the boat, we came upon the entrance to the tunnel at Les Cammazes. One of Riquet's great innovations on the canal was to build a tunnel, and to build it, moreover, against a tide of popular opposition. Nobody had ever had the idea of digging a tunnel through the rock and then filling it with water – such a notion was surely perilous as well as absurd.

This statue of Riquet stands as a proud monument to his work, watching over the main square in Béziers.

Work on the tunnel was ordered to cease. But Riquet, who was following the evident example of tenth-century monks who had penetrated part of the surrounding land as a means of drawing water – and who therefore knew his idea was feasible – persevered. He must have enjoyed a good relationship with his miners, who, in their determination to succeed, worked at night to elude the inspectors who had been posted to make sure excavation had been abandoned.

One more unusual engineering feature made us get out and explore. At one point, the canal crosses a river. Now the normal thing to do would be to construct an aqueduct for the higher stream over the lower (this might have to be assisted with locks). But the trouble here was that canal and river were at exactly the same level, so a system called an *ouvrage*, on the principle of a railway level crossing, has been installed to allow the river water to flow freely into the canal unless it is locked off, to allow the passage of boats and to control floodwater.

At Agde, there is an unusual three-way lock, one path leading into the town itself, the other following the canal on its short final journey into the Étang de Thau, and so to the Mediterranean and farewell.

Pierre-Paul Riquet never lived to see the completion of his life's ambition, nor did he see the ceremony held to mark the opening. Bishops, parliamentarians and other notables, Riquet's family members, a band and a vast amount of food were drawn all the way from Toulouse into Béziers, a journey that lasted ten days. In the rear of the procession were thirteen barges laden with French, Dutch and English goods destined for the fair at Beaucaire; superintended by Cardinal Archbishop Bonzi.

In 1845, Hans Christian Andersen had taken the journey by boat from Toulouse to Sète. He writes:

> Beneath in the cabin it was crowded; people sat close together like flies in a cup of sugar. I made myself a way out through boxes, people and umbrellas, and stood in a boiling hot air; on either side the prospect was eternally the same: green grass, a green tree, flood-gates; green grass, green tree, flood-gates, it was enough to drive a man insane. The sun burnt infernally. People say the south of France is a portion of paradise: under the present circumstances it seemed to me a portion of hell.

Well, you can't please everyone. Pru and I enjoyed a wonderful journey, and we're very keen to come back; perhaps see a little more of the canal at the other, western, end. Ideally we would like to wait a year or two until the plane trees, famously shading the banks of the Canal du Midi, have grown up to replace their poor diseased forefathers. We'll see.

THE
LONDON RING
CANALS

FOR OUR NEXT PROGRAMME, we had been thinking about
that group of little-navigated canals threading through the
heart of London itself. We didn't know much about them;
of course, we'd crossed over the Regent's Canal many times
on our way up to Swiss Cottage and Hampstead; we'd
driven past Little Venice, and fought our way through the
crowds around Camden Lock. We also knew Battlebridge
Basin, because long ago we thought we'd seek a permanent
mooring for our boat that was easily accessible from home.
That would have given us a perfect opportunity to study
the London waterscape, but in those days we were more
interested in how long it would take us to get to Greenford
so that we could finally see a bit of country. So that didn't
last, and the following year we left the Basin and went back
to Newbury.

But, apart from these separately recalled landmarks, the

various bits of waterway that served the metropolis and disgorged its produce into the Thames at Limehouse, again at Brentford, and northwards on to the Lee Navigation, remained an enticing mystery, even though so close to home.

We started off at the back door of the Grand Union Canal, as it sprouts off the Thames at Brentford, opposite Kew Gardens. In the Canal's early days, a lot of boatbuilding and repairing, as well as shipping goods from canal to Thames barges and vice versa, went on in the few miles going westward until it meets the Paddington Arm at Bull's Bridge Junction.

Apart from our film crew, we also travelled with a 'skipper', representing the boat owner, who was on hand to sort out any technical problems and could reposition the boat to another site while we were filming ashore. We also chartered a 'relief boat' that could weave its way round our own vessel and film us from all angles. On shore we had two vans with two regular drivers, to carry us about and do the shopping. That was everybody.

A word about filming in narrowboats. The camera crew always travel with a great deal of stuff: tripods, lens bags, battery bags, things to sit or kneel on, slings to keep you from falling off things. The sound department seem to have a lot of extra luggage, too. All this has to be stored somewhere: on the beds, so that you can't lie on them; in the shower, so that you can't wash; in the lavatory, demanding intricate bodily adjustment.

Added to this is a visibility problem. While Pru and I are usually out on deck, down below, there may be our director, his PA, the sound recordist, one of the two cameramen, the boat's 'skipper', and somebody making lunch. While we're being filmed from the shore, or from our relief boat, none of these people must be seen through the windows: Pru and I are supposed to be alone on the boat. This means a lot of these people spend a great deal of time lying on the floor, clutching walkie-talkies that will tell them when it's safe to get up.

I'm afraid I got impatient with Pru over the filling of our first rather deep lock, and I must watch this tendency. Her hearing has deteriorated along with her general condition, and I get angry with myself for not taking this into account. I suppose it's natural that the longer you've known someone,

Locks can be stiff work when one is in one's mid-eighties.

147

the more you take it for granted that they can't change substantially; but of course they can, and do. Darling Pru is getting a little less sure of herself all the time; only a little, perhaps, but I ought to be constantly on the lookout. She insists on doing tricky things, because she loves it, but of course needs constant encouragement; so getting irritable with her is the worst thing I can do. That was Clitheroe's Lock, 7 foot 7 inches deep, so the tide rushes in above water level and you have to control it.

At the six Hanwell Locks, however, we had two sturdy CRT lock-keepers to help us up the climb. At the top is an aqueduct known as The Three Bridges, because it crosses over both a road and a railway track at the same time. Isambard Kingdom Brunel was responsible, and it is thought that this may have been the very last work he undertook. We enjoyed looking down on the busy South Circular from our stately progress at four miles an hour.

While of course I'd known about Kensal Green Cemetery, and indeed visited it once or twice, I'd always come by road, and never geographically connected it with the canal. But now here it was, on the port side, with a little over-grown gate leading onto it from the canal bank. We opened it and it felt like coming out of the TARDIS. Just off the Harrow Road lies this peaceful seventy-two-acre space, which is a wildlife sanctuary as well as the historical burial ground modelled on the Cemetery of Père Lachaise in Paris. A haven of extraordinary ease and serenity, we wandered about for some time reading inscriptions, before returning to the boat and the much less tranquil environment of Paddington Basin.

Where to moor for the night? Turn right before you get to Little Venice, they said, and there are supposed to be comfortable moorings. So there are, but they're all taken. We ended up under the M40.

The basin, with its proximity to Paddington Station, was naturally once crowded with wharves and warehouses, then it gradually fell into disuse and decay. But in 1999, in line with much top-class canalside evolution, the basin was drained and redeveloped.

As you pass along the canal admiring some of the elegant and striking commercial buildings that are taking advantage of a waterside aspect, you sometimes wonder a bit about social housing? Not much evidence.

The London locks: occasionally less than beautiful, never less than bustling.

In Little Venice, the stretch of the canal so christened by Robert Browning, we picked up Andrew and Melody Sachs. We'd each of us known Andy for many years before his immortal Manuel on *Fawlty Towers* (apparently, when it is shown in Spain, Manuel has to be an Italian) and shortly after our meeting we worked together in the last thing he ever did: our characters were in the same ward of a geriatric hospital in *EastEnders*, hurling insults at each other across the room. His character died, I'm afraid; and, not so very long after, so did Andrew himself. He was a man of many different talents, of which perhaps not enough people were aware.

The canal calls itself the Regent's Canal for the rest of its length, skirting John Nash's Regent's Park and Regent's Park Zoo. The zoo has been there for 188 years in this wonderful situation, but faces the problem of fulfilling an obligation to expand, in terms of both space and zoological representation, and has no way of extending the present site.

It was a warm September Sunday, and crowds of people were picnicking along the bank around Kentish Town. Perhaps the word 'picnic' gives the wrong impression: it wasn't cucumber sandwiches and tea from the thermos, but slices of pizza and tins of Red Bull. And something else, I'm sorry to say. Seeing us approaching with a film camera on board, one man (it turned out he was an actor; perhaps he reasonably objected to being filmed and not paid for it) grew quite abusive. There were some Words between him and James (who wasn't shooting in this guy's direction anyway), before two police officers appeared from nowhere; quiet reigned, substances were hastily concealed, and we continued on our way.

It might, I think, have got rather ugly, and I was a bit nervous for Pru, standing there on the foredeck, but she was completely unabashed; in fact, I think she rather enjoyed it. Good old Pru!

We turned off to revisit Battlebridge Basin, which now houses the London Canal Museum, and is a square stretch of water between four walls of sedate, nineteenth-century, converted warehouses. It was very quiet, very calm, and once again one had to remind oneself of a canal's ability to exist in phlegmatic solitude within a stone's throw of bustling Islington life. Martin, the museum's director, showed us some photographs of how things must have looked here at the beginning of the twentieth century: boatmen and their families unloading cargoes of ice, transhipped from Norway and brought up the canal from Limehouse.

It had never occurred to me that ice was an important commodity worth that amount of trouble and expense; but, of course, in hot weather food storage had to be surrounded by ice. People excavated under their houses to create ice houses among the rocks. A flourishing and profitable business must have been brought to an end by the inventor of the refrigerator.

We went through the Islington Tunnel – at least cameraman Gary and I did, while Pru walked over the top with James and Sam and did a lot of lusciously edible shopping in the Islington lanes. (This is where Sam and his partner Laura live now, and I can quite see why.)

The tunnel was built in 1880, with 4 million bricks; no

Inspecting 4 million bricks for structural safety takes a bit of doing. . . © Getty

towpath, so when it opened the traffic must have been steam-powered. A bit further on, we came to the short Hertford Union Canal, linking us eastwards to the River Lee. We decided we'd go down it. It is a very sharp, very narrow turn, and not achieved without a good bump to satisfy Channel 4.

It was an extremely unsuccessful project, opened in 1831 and used very sparingly. Strangely, though, it has survived and does offer a pleasing view right across Victoria Park, which it skirts on its left side for most of its length.

We turned south onto the Lee Navigation. By now it was getting dark, and I really had no idea where we were planning to moor for the night. Downstairs, Mike and Trina were busily ringing round (apparently, we were due at some place, but it had been double-booked) and getting nowhere. This was an area – Sweetwater, Marshgate, Pudding Mill – that I had never actually heard of, let alone visited by other means. I plunged on. Weirdly lit underpass after underpass, strangely beautiful, but, in the otherwise darkness, giving no hint of a location.

Suddenly, Mike was beside me. 'Pull in here.' I could just make out a landing stage ahead, with enough room to moor behind another boat, which I did. The other boat turned out to be a travelling sixteen-seat cinema run by two enterprising girls who took us on board and offered to run a film for us. I'm afraid we were all too tired. Pru and I went to bed, and somehow the crew got hold of a taxi back to their hotel.

When we woke up, the girls and their cinema had gone. It was a lovely morning, no one about, and the scene looked like a late-seventeenth-century painting. Without knowing where we were, we'd tied up outside the famous Three Mills – the last survivors of a group of tidal mills on the Lee, used mostly to grind maize and barley for distilling purposes. To most people now, I suppose, the name suggests Three Mills Studios, which are on a little channel round the back and are a thriving film and TV resource.

The crew arrived, and we pressed on down the Limehouse Cut to the Limehouse Basin marina. Pru and I walked down to the river and turned into Narrow Street, where there's a pub called The Grapes. Not only is it a very good pub, but it is owned by our friend Sir Ian McKellen. Ian was there,

The Grapes is rather a nice spot for a pint, and the company wasn't bad, either.

and we had a chat about the pub and its history, and drank a nice glass of beer, and then returned to our boat.

We were in for a shock. It appeared that a few days ago a narrowboat had ventured on to the Thames and had got into some sort of trouble: summarily, there were no more narrowboats allowed onto the Lower Thames. What could we do? The whole point of the programme was to complete the circuit clockwise from Brentford to Brentford, contrasting the two routes.

We took our dilemma to the skipper of a well-seasoned-looking motor launch moored up in the basin, and for a consideration he said he'd take us upriver to where we started.

The ride was quite choppy, and we were rather relieved not to be doing it in the narrowboat. We sped along, and were given an unlooked-for treat in passing the beautiful tall ship *Stavros Niarchos*, her hands standing up on the yards, making her way down the river and out to sea.

At the end of this trip I thought we really ought to do something about Pru's increasing hearing difficulties, so we sought some advice, and settled on some really state-of-the-art devices. The only trouble is, she hasn't really accustomed herself to having objects in her ears, and either forgets, or is reluctant, to put them in.

This morning:

Pru: Have you seen my hearing aids?

Me: No I haven't, sorry. Where might you have put them?

Pru: Sorry?

Me: I said, where do you think you've put them?

Pru: I can't hear you.

Me: No, I know you can't, because you haven't got them in.

Pru: No, I'm so sorry. Where are they?

Me: Well, when they're not in your ears, where do you leave them?

Pru: In this box here.

Me: Well, they're not there now, are they?

Pru: Sorry?

Me: Why aren't they in your ears?

Pru: I take them out to wash my face.

Me: I see. So then where do you put them?

Pru: Sorry?

Me: *Where do you put them when you're washing your face?*

Pru: In the box. Oh, wait a minute. I think there's another box.

Me: Another box. What does it say?

Pru: Hairpins. Here it is, there they are. I'm putting them in. *Hooray!* Now I can hear you perfectly. What did you say?

Me: I didn't say anything.

Pru: Sorry?

I'm sure the hearing aids are very good; the thing is, I'm not quite sure that it's an entirely physical problem. Sometimes people say things to Pru that it takes her a moment to comprehend; and I just wonder if she's saying to herself, Ooh, I need a bit of time here; shall I pretend I haven't heard?

I think she does it with me. The effect it has on *me*, of course, is that I seldom feel that what I've said is worth repeating for a second or third time, so conversation falters. But we're learning to communicate without language.

INTERLUDE
THE FAMILY BUSINESS

Pru: Where's my . . .? [pause] I can't find my . . . [pause] It's *always* here. Where is it? [calling] Timmy!

Me: Hello?

Pru: Hello!

Me: What do you want?

Pru: I can't find my . . . Have you seen it?

Me: Have I seen what?

Pru: What?

Me: What is it you want to know whether I've seen? Or not?

Pru: I'm sorry, I can't hear you. I'm coming up.

[The telephone rings. Pru gets there first.]

Pru: Hello? Oh! How lovely to hear you!

[There's a long animated conversation, before she finally puts the phone down.]

Me: Who was that?

Pru: Oh, you know, whatsername, from across the Common.

Me: No, I *don't* know. Who?

Pru: Well, no, actually, I don't think you do know her.

Me: What did she want?

Pru: Sorry?

Me: What did she *want*?

Pru: Oh – I don't know, nothing really; I think she just wanted a chat. [Pause] What did I come up here for?

I decided to call 1471 to find out who the lady was, and Pru was quite right: I didn't know her; and, yes, indeed, she had just rung because she was lonely and wanted someone to talk to. This is something Pru is amazingly good at: comforting and reassuring people who are ill, perhaps bereaved, or in need of encouragement. It's a very precious skill but it's not a skill at all: she really does empathise and share in the sorrow. And the astonishing thing to me is, she can do it on the *telephone.*

Not like me. I have always regarded Mr Bell's instrument, in whatever state-of-the-art form it comes, as a fundamental deterrent to human dialogue. When I'm on the phone I hear myself saying things I don't mean, and completely misunderstanding whatever is said back to me.

Worst of all, when the telephone rings I am immediately overcome with feelings of *guilt*: the call must surely be from someone I've failed, let down or forgotten all about.

I remember this happening years ago. I picked up the phone and a stern female voice announced: 'This is Lady Margaret Hall.'

Oh, God! I thought; *yes*, I remember, this is about some

event I was meant to be attending, some appeal to which I'd said I'd respond, some letter I was going to write in support of something – and *I haven't done it.*

'Hello, Lady Margaret,' I gabbled. 'Look, I'm sorry, your letter is on my desk. I haven't yet had a moment to attend to it—'

A tired sigh came down the phone. 'No, this is Lady Margaret Hall, Oxford. I am just ringing to tell you that your son Samuel has been accepted for a place in this college, starting next term in September.' Her tone suggested that, if his father's intellectual powers had been passed on, Samuel's stay in Oxford would probably be quite short.

Sam had rather a good time at LMH. He did a great deal of acting, some directing, toured East Africa for the Oxford University Dramatic Society (or OUDS) in the summer break, and came away with quite a respectable degree.

We are both very proud that he went into the family business, and is doing so very well at it. As well as being an outstanding actor, he has great talents as a director: clear, imaginative and truthful, and with intent to serve the playwright above all else. People love working with him.

He did a spell as artistic director of the combined Sheffield Theatres. An artistic director has three jobs – well, four, really: running the company, running the building(s), directing some plays and being the elected cultural mouthpiece for the community in which he or she works. You have to be in sympathetic partnership with your administrative director, and earn the confidence of the board of trustees.

Sam performed alongside his mother in the film adaptation of E. M. Forster's masterpiece, *Howard's End*. We are incredibly proud.

Halfway through Sam's term of office, it was decided to do some serious and necessary structural work on the Crucible, the city's main producing theatre. It would have to close, for probably eighteen months. What would happen to Sam's programme of future productions? What would happen to the creative unit of the company? What would happen to the stage staff: carpenters, scene-painters, electricians?

Sam went all round Sheffield, seeking alternative temporary accommodation, and had some success. The

Cathedral wanted to house a production; a musical had been written about the decline of the Sheffield steel industry, and one of the old foundries could be made available. Even Sheffield Wednesday offered a bit of space. Sam brought all these ideas to the table, but was met with a negative response. It was all too risky, too difficult. Let's just close, they said; bank the annual grant, lay off the workforce and meet again in eighteen months.

That's what happened – but without Sam, of course: you don't need an artistic director if you've no art to direct. I wonder if he'll try again, somewhere; I hope he does.

I've done it myself a couple of times. In 1973, I was asked if I'd like to form a company to do some plays at the Forum Theatre Billingham, on Teesside. Billingham's a strange place: no canals, of course, but ICI chemical plants everywhere, and in the middle a vast building called the Forum, comprising a huge swimming pool, an ice rink, squash and badminton courts, boxing rings, cafeterias and bars. Oh, and a theatre. It was supposed to be a meeting place for the whole community, where all-in wrestlers would sit down with patrons who had just seen a performance of *Ariadne auf Naxos* by Opera North, and compare their evenings over a companionable drink. It didn't quite happen.

Curiously, though, it gave me licence to put on large-cast plays with big, solidly built sets. ICI paid enormous rates to the Forum, and the theatre had a workshop crammed with timber, presided over by a production manager and two carpenters I didn't have to pay for. There seemed to be scarcely a limit to my budget: at one time I was employing thirty-two actors, including a theatre-in-education team.

We did modestly all right for audiences, but not from the immediate surrounds of Billingham: they came from classier Durham, from Bishop Auckland and Stockton-on-Tees. The owner of the local petrol station told me he'd really like to come and see a play, but he couldn't afford the time off to wash and put on a suit. I told him that wasn't at all necessary, and that Myra really didn't need to go and have her hair done; but they didn't believe me.

There was a theatre board, but few of its members came to see a performance; in fact, one of them had never set foot in the auditorium because of his fear that the neighbouring swimming pool would somehow overflow and drown him. I produced the architect's working drawings, showing clearly the water level in the pool to be well below any part of the theatre, but he wasn't convinced. 'I'm not taking the chance,' he said firmly.

My second (and final) foray into theatre management was when I had to take over the artistic directorship of the resident company at the Old Vic in London. Here, the financial situation was very different. The historic and lovely theatre was in a bad way structurally: the electric cabling was very nearly actively dangerous; a lot of the seating was very uncomfortable; and the carpets were patched with unsightly gaffer tape.

My position was that of the cricketing nightwatchman: I had somehow to protect the wicket until relief came in the morning. It didn't come. It was a time of savage national cuts to artistic institutions, and the Arts Council withdrew our entire grant in 1981.

Fortunately, help was at hand in the person of Ed Mirvish, the Canadian impresario, who bought the theatre, did all the repairs and invited in a series of individual productions until Kevin Spacey came over from the States to take over the artistic governance of the theatre, and returned it triumphantly to being a recognisable location on the London map. For years, there had been a theatregoing reluctance to venture across the river – despite the fact that there is an enormous railway station at Waterloo, served also by three Underground lines and numerous bus services. On the other hand, people somehow don't seem to mind diving down into an insalubrious cardboard-city subway to emerge eventually in the skateboarding environs of the National Theatre.

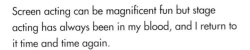

Screen acting can be magnificent fun but stage acting has always been in my blood, and I return to it time and time again.

PART 4

LONDON'S LOST ROUTE TO SEA

EVER SINCE I HEARD about it, I've been intrigued by what has been called London's Lost Route to the Sea.

From early in the seventeenth century, different schemes have been put forward to find a way from the metropolis to the south coast without having to take heavily laden ships down the Thames Estuary and round East Kent to face the elements in the Strait of Dover. Finally, the route that appeared most sensible was to find a point in Sussex where two useful existing rivers could be united: the Wey, flowing north to join the Thames at Weybridge, and the Arun, going in the opposite direction to emerge into the Channel at Littlehampton. At their closest point, the rivers lay only ten miles apart, navigable most of the way, and yet it took 175 years for the Wey and Arun Junction Canal Bill – an immensely popular project – to get its first reading.

During the Napoleonic Wars, providing money and

provisions to our fleet patrolling the English Channel became a safety problem: the Kent coast was inundated with pirates in addition to the hostile French. Now that the Wey and Arun was actually a going concern, there were enthusiastic demands to build a link, from below Arundel through directly to Portsmouth, so that gold bullion could be brought inland from London to be paid out to our sailors in fortified safety.

As we know, Napoleon was unsportsmanlike enough finally to lose the war; so the scheme was no longer necessary and the connecting arm to Portsmouth never did get built, although traces of it can still be seen along the way. With Portsmouth no longer being viable as a destination, enthusiasm waned for the whole route, and sections of it gradually collapsed through lack of use.

When we decided to make our journey in 2014, however, we reckoned that about 85 per cent of the passage was just still navigable. We would have to take the road for a short while, from just below Shalford to the village of Loxwood; and then, after a four-mile reclaimed stretch of canal, get back in the car to drive to the point where the canal joins the River Arun. After that we were fine.

In my pocket was a little book called *The Thames to the Solent by Canal and Sea, or the log of the Una-boat 'Caprice'*, by J B Dashwood. In 1867, Mr Dashwood with his wife and dog (I don't know why we never referred to the dog in our programme) set out from Weybridge in the sailing dinghy *Caprice* to navigate the waterways through to Portsmouth to witness the Naval Review at Spithead.

For our attempt to re-live Mr Dashwood's adventure,

a chartered narrowboat was waiting for us just above Teddington Lock, the furthest downstream lock on the Thames, and therefore the limit of the tidal river. Next to it is Teddington Studios, once the headquarters of Thames Television, who used to broadcast programmes in the London area from Monday to Friday; on Saturday and Sunday it was the turn of London Weekend Television. The Thames studios were a pleasure to work in: a quiet road with a nice pub next door, and a canteen where you could sit and watch the passage of boats through the lock.

We went up the river, past Hampton Court to Tagg's Island, originally Walnut Tree Island, populated in the mid-1850s by a number of squatter families who made

Garrick Temple: the shrine within which a superstitious traveller can deliver their offer of a poem to the spirit of William Shakespeare, in return for a safe journey.

a living out of cutting willow rods, which they peeled, bleached and made into baskets. The island has a rather different look today as a mooring for wealthy boat owners, who have in the past included the showman/impresario Fred Karno and the playwright J M Barrie, author of *Peter Pan*. Karno is popularly credited with discovering Charlie Chaplin, Stan Laurel and Oliver Hardy.

We were now in Sunbury, and here our old friend and colleague Clive Francis invited us into Garrick's Temple, a small eighteenth-century circular building erected by the great actor David Garrick as a monument to Shakespeare. Clive is chair of the trust that owns the building. He gave us tea and asked me to compose a short poem to the Bard, which is traditionally required to ensure his blessing for our voyage.

All actors should be conscious of their debt to David Garrick, who introduced to the theatre a new naturalism, taking over from the formal declamatory style to which audiences had been accustomed.

Sometimes, even today, it is necessary to remind people. As we were saying goodbye to Clive, Pru remembered a story about our son Sam at his school, Alleyn's, where the tradition of school drama seemed to depend on standing face front, in a lot of makeup, and shouting.

Sam thought this really wouldn't do, and got together with some friends to put together a studio production, in their own time, of Max Frisch's play *The Fire Raisers*. They were rehearsing after school one evening when the headmaster and another teacher came through the room talking loudly and went out of the other door.

'Excuse me, Headmaster,' said his colleague. 'I think you may have interrupted a rehearsal going on back there.'

The head went back to apologise. 'I'm sorry, Sam. I didn't realise you were rehearsing. I thought you were just talking to each other.'

'Well, sir,' explained Sam, 'that's what it's *meant* to sound like.'

(I have to say that drama at Alleyn's these days is of a very high standard, possibly because half the parents I meet there seem to be in the business.)

Garrick's blessing guaranteed our safety as far as Weybridge, where we spent the night. In June 1940 this junction with the River Wey was selected as a muster point for all the 'little ships' commandeered to take part in the massive rescue operation of our troops at Dunkirk. Some of these little craft, as we know, were tiny, safely capable of carrying only perhaps half a dozen people, but thousands owe their lives to the skill and courage of those civilian boat owners who set sail down the Thames that June morning. Perhaps they would have benefited from the alternative Route to the Sea.

We branched off onto the River Wey. I knew something about the river, passing as it does the Yvonne Arnaud Theatre at Guildford, with a lock that is almost adjacent to the stage door. In common with most people, I had always considered Guildford as a successfully retired market town whose well-heeled population enjoyed the access to the Surrey countryside and the excellent train service to Waterloo. Approaching the town by water, however, reveals Guildford's very different past: a busy agricultural trading-post with wharves and warehouses still lining the banks.

The Yvonne Arnaud is a theatre of which we are both very fond, and it was lovely to be welcomed ashore for a glass of champagne. Pru and I are both very committed to touring theatre; we feel that the rest of the country deserves the same access to good drama as is available in the metropolis. For us, too, it is good to feel you've been allowed to feel part of a community for a week, and to play to people in the evening that you're going to meet in Sainsbury's in the morning. That doesn't happen in London.

Guildford's theatre management have always been resolute in producing, or helping to produce, quality work

A very civilised welcome to the Yvonne Arnaud Theatre.

that will then tour the regions. There is a chance that the show may be picked up for a West End transfer – if it does, fine, and it will refund the production costs and make a bit of profit; but, if it doesn't, still the play will have got an airing and been seen by a lot of people all over the country.

Soon we were at the village of Shalford. The Wey continues on to Godalming, a stretch that continued to be navigable until 1950; but we turned off onto the beginning of the Wey and Arun Canal. It doesn't go very far, and comes to a halt at a low bridge under which Mr Dashwood, by lowering his mast, apparently ducked and sailed on; but we found energetic members of the Canal Restoration Team who told us that hard work is being undertaken to ensure the whole route to the Arun will, eventually, be open.

A branch of the campaign has already made progress, and opened a section from its headquarters at Loxwood to a point about four miles further to the east, and there we were invited to explore this stretch in a special little boat more adapted to the reclaimed waters, which are shallow, slender and very beautiful.

Before we did that, though, we took a walk away from the canal to visit The Old Forge at Abinger, the house where Pru was born. We met the present owners and their two little girls, who showed us round. It must be a strange experience to look upon the actual room in which you took your first breath. An experience not open to me, I'm afraid: I was born in a house in Bradford, which at the time was a nursing home, and doesn't bear any traces.

Another fairly short car journey delivered us to where the third boat on our pilgrimage would take us downriver to

Arundel Castle, where archivist Dr John Robinson told us more about the Duke of Norfolk's interested involvement with the plans to link London to Portsmouth by canal. Finally, the Arundel harbourmaster give us a lift on the final leg to Littlehampton and the sea, passing on our right the derelict entrance to the Portsmouth tributary that Mr and Mrs Dashwood hoped would have been functioning in time to carry them to witness the Royal Naval Review.

Of course it wasn't. That must have been a disappointment. Nonetheless, they must have felt gratified at having completed the whole journey from Weybridge to Littlehampton in one very small boat. It took us four different craft and two motorcars to achieve the same thing; but, still, achieve it we did, and we all felt very proud of ourselves.

At the Old Forge, the cottage where it all began for Pru.

177

BIRMINGHAM
TO BRAUNSTON

PERHAPS WITH A CHILDHOOD like mine, a disposition
to 'itchy feet' is understandable. I've always loved the idea
of moving on somewhere else, seeing some new things.

Well, if that's what turns you on, people say, why don't
you do it in style: take off on a cruise liner?

Well, we've done it, a bit; the trouble is that I'm so
fond of ships that I can't bear to go to sea in something
that doesn't *look* like a ship. We did once find a vessel
that had a recognisable maritime shape; she was quite
small, accommodating about 550 passengers and was
quietly comfortable. Consequently, the owners went out of
business, and we haven't ventured again.

Pru feels guilty about being looked after all the time:
regular food, cocktails, planned shore visits. I think I could
live with all that; what depresses me is what cruise-ship
culture is doing to the world. Every time I stepped ashore

in a new port, I used to ask what the local industry was. I always got the same cheerful reply: tourism. Places that had once been attractive little villages or fishing harbours are given over to bars, tourist shops and fast-food restaurants. And they are all beginning to look the same: Greece, Turkey, Italy, Spain. With a hotel ship in the harbour, and two more on the horizon waiting their turn. And a huge one parked, daily, outside San Marco in Venice.

If we were younger, we'd embark on a sailing ship; one of those where you're expected to climb the rigging and *do* things. Pru has always had a love affair with square-rigged ships; I don't know how it started. But, no, we're no longer of an age. Better stick to canals.

So. Our next idea for a *GCJ* programme was to link together Birmingham and Braunston, two centres of essential canal history, along the Grand Union Canal. Birmingham, with its network of industrial canals, is described by our son Sam, who joined us on the trip, as the Venice of the Midlands.

We started off in Gas Street Basin, which still has a whiff of the bygone era of historic working boats, in spite of having been smartened and prettified. In the 1980s, I filmed here, at night, for a TV thriller series. The area then was extremely sinister, and the water thick with oil and rubbish. Our leading actor had to be thrown into the canal, and the insurers insisted on his taking an emetic before he went in, in order to sick up anything he had accidentally drunk. I remember he had to do three takes: three emetics, three vomits. Television was hard work in those days.

Nowadays the whole area, named Brindley Place after the engineer-architect, is resplendent with restaurants, bars and little shops, and an attractive walkway into Birmingham's splendid Symphony Hall.

Travelling east out of Birmingham, you've eventually got to face the arduous Hatton flight of locks. By a wonderfully lucky chance we found a crew of CRT maintenance men who were about to work their way down the twenty-one-lock staircase, and they agreed to travel with us and see if we could break the track record of, some say, two hours and five minutes. We did it in two hours seven.

We'd left our own boat for a few weeks at the nearby

On this journey, we met the remarkable Sarah Henshaw on her Book Barge, the amazing floating bookshop she patrols Britain's waterways in.

Calcutt Marina, so we swapped over in order to get it back on to our official mooring at Braunston. It was good to be back on our own boat. During the series, we like to sleep on board the boat that has been provided for us whenever that's possible; but very often it isn't, and it becomes a series of hotels – usually a different one every night, because filming locations are so far apart.

In general, I like staying in hotels (good ones, anyway), but this is not the best way to enjoy them. At the end of each day's shooting, we meet the unit for dinner somewhere before getting back to our hotel room, unpacking and sorting out what clothes we'll need to wear for tomorrow. No, you

Pru always did make friends rather easily, and this stop at Heronfield Animal Rescue Centre in Solihull proved no exception. I must admit, it is a very winsome alpaca.

can't wear that shirt again, you need the blue short-sleeved one as continuity from Wednesday. But you can have the same trousers. No, I can't, they're filthy. Well, you should have brought a spare pair. Won't the cream-coloured ones do? No, they won't match what we did yesterday. And what if it's raining?

At seven in the morning, my wonderful daughter Juliet, who as a professional hairdresser has been retained to look after Pru's hair and makeup, knocks on our door to get us up. She makes tea for us and gets to work on Pru: I lie in bed till I can get into the bathroom. We repack, grab a snatched breakfast downstairs and, at 8.30, the van comes to drive us to today's location, wherever that may be; while Trina, in the back seat, explains all the last-minute changes of plan that have become necessary since we spoke last night.

This morning, at Braunston, we know exactly what we have to do: this is the day of the great national Narrowboat Rally, and Pru and I have been given the task of leading the parade of historic working boats that will leave the boatyard, followed by scores of privately owned craft, process along the canal, turn at the junction with the North Oxford, and make their proud way back into the Marina. I was to be aboard *President*, the last remaining *steam-driven* working boat on the system, built in 1909 to maintain for fifty-four years a nonstop cargo service between Birmingham and London.

So today Pru and I are in double service: to the television programme on one hand, and on the other to Tim Coghlan, the impresario of the day's proceedings. First of all, Tim had provided us with period boatman's and boatwoman's outfits to wear in the procession. Now, actually, the thing I enjoy least about getting into character is putting on the costume. People find this strange; Tim Coghlan does. 'Oh, but I thought you people *liked* dressing up,' he said to me. No, Tim. The shirt scratches, the trousers are too long and the immensely heavy belt doesn't fit into the loops. But, even when a costume fits perfectly and even looks right, it doesn't feel part of me until I've worn it for a few days, doing the sorts of things it was designed for.

Pru, on the other hand, looked a picture in her bonnet, bodice and striped shirt. It was a warm, sunny day; there were no mishaps during the voyage; the beer tent did amazing business; and in the evening there was a show performed by the Mikron Theatre Company, of which we are patrons, and from whose offices in Marsden outside Huddersfield they tour the country's canals from May to October, giving performances at Town Halls, schools,

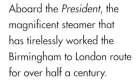

Aboard the *President*, the magnificent steamer that has tirelessly worked the Birmingham to London route for over half a century.

cricket pavilions and waterside pubs. The show we saw was excellent, and packed out; drawing not only the Rally audience but the locals from up the hill.

Braunston village still sees itself as belonging to the canal, even though it is situated up a steep rise away from the water. In the village churchyard rest the bones of working boatmen from the nineteenth century, for whom Braunston finally supplied a home address. Although they were an occupation exempt from enlistment during the first war, five of the names on the village roll of honour have been identified from the boating fraternity.

SHANNON-ERNE
WATERWAY

UP TO NOW our series had involved itself quite a lot with Pru's background: we'd looked into her Yorkshire forbears, her French romance (followed by her Scottish romance) and her Surrey childhood, and we'd finally visited the house where she was born.

I think there may have been an attempt to redress the balance in my favour, but nobody came up with any ideas until, while we were discussing a notion of mine to see the canals of Ireland, I let slip that I am in fact half Irish.

So we decided we should do the trip, but what they didn't tell me was that they had engaged the services of a genealogist to meet me and sort out my background. So I left England thinking of myself simply as a member of an averagely competent theatrical family, but I came home the apparent descendant of an aristocratic Anglo-Irish bloodline, one of whom had served as Governor-General of Canada, no less. It shows you how following a canal can often lead you into surprising waters.

I am half Irish through my mother. As a child I used to accompany her on her occasional visits to County Wicklow, but, apart from a few professional visits to Dublin and to Wexford over the years, I had not been back to the Old Country. Does the Emerald Isle include Northern Ireland? I hope so: I have performed in Belfast a few times, and filmed in the area, and one of the prospects that excited me about exploring the canals here was to be able to cross from South to North on 'water that knows no boundaries'.

The history of waterways in the Irish Republic is a rather unhappy one. Two major canals survived the general collapse of commercial traffic: the Grand and its rival the Royal, both leading from Dublin to the River Shannon by different routes. Each experienced physical and economic problems during the twentieth century, although programmes of restoration work are now in hand. Other Irish canals were not so fortunate.

The Shannon–Erne Waterway reopened in 1994 to link the two rivers Shannon and Erne. It is really a series of lakes connected by other rivers and the odd bit of canal, and it was this route we decided to pursue. In our curiously shaped narrowboat *Maeve* (should we be saying 'narrowboat'? – she's a comfortable 8 foot 6 inches in breadth), we followed the Loch Erne Canal, which has the reputation of being the least successful canal in commercial history, and crossed the invisible line from Northern Ireland to the Republic, passing the ruins of a bridge that had been blown up during the conflict, and has now been replaced by a new bridge financed by funds from the Peace Process.

There is a series of locks along the canalised section of our route, mechanised and operated by the insertion of a small plastic card, which perplexed Pru, a windlass operator of the old school, until she got the hang of it.

Frank McCabe, the lengthsman (responsible for lengths of towpath, as well as the locks in the absence of a lock-keeper), joined us aboard and told us how the canal construction was hampered by starvation in the terrible potato famine of the 1840s. A million died, and a further million emigrated abroad, mostly to America.

When one is presented with a fresh pint of Guinness, one simply does one's best. . .

It rained solidly throughout our first day, and in the evening we were glad to pull in at McCarthy's traditional pub in Leitrim, where they got us singing and dancing, and drinking quite a lot of Guinness.

That evening we were visited by a *seanchaí*, a professional storyteller dealing in Irish folk legends and stories that have been orally handed down through the generations. Pru wanted to know about the 'fairies' that are so often talked about, and it was explained that the local variety are in no way related to the delicate little winged creatures so beloved by Victorian illustrators. No, traditional Irish fairies were indistinguishable from you and me; they just had special powers, which included the stealing away of children (*boy* children only) to swell their ranks.

This apparently explains why, in some early black-and-white photographs of rural Irish families, small boys are occasionally dressed up as little girls. To fool the fairies.

As the evening skies closed over the boat's cabin, we lit a candle and our *seanchaí* produced a bottle of whiskey. As we supped, our discourse turned to ghost stories. Had we ever had experiences of the supernatural? we were asked.

There has just been one occasion when I felt in the presence of a ghostly manifestation – not a single human presence, but a community from the past. I was staying with some friends in a very old manor house in Warwickshire. I was performing in a play in Birmingham, and by the time I got home at night it was quite late and my hosts had gone to bed. They had left a key to the front door where I could find it, and as I picked it up I felt an extraordinary blast of cold air coming from behind me. I couldn't understand it:

it was a warm summer's night, and behind me simply lay open fields, nothing else.

As I looked into the distance, I was faintly aware of some tiny lights flickering away on the horizon, where the force of cold air seemed to be coming from, and, with it, what seemed a powerful current of *hatred.* Yes. I was being told I really was not wanted here. I lost no time in opening the heavy oak door and shutting it safely behind me.

At breakfast next morning, I told my hostess of my experience, and she, not greatly surprised, told me the whole story. Where I had seen the phantom lights last night,

Some of the waters along the Lower Lough Erne are very atmospheric – conduicive, perhaps, to supernatural experiences like mine.

191

there used to be a village, which fell a victim of the Great Plague of 1665. The Lord of the Manor (where I now was) gave orders for the village to be burned, to stop the spread of the infection around the countryside. Apparently, his instruction was accepted philosophically by the villagers, but what really upset them was his removal of the great oak door from the church and installing it at the front of the Manor, which had remained unaffected by the plague.

I'm afraid I couldn't pretend to be the first to witness this phenomenon: others had had similar experiences while staying in the house.

We said goodbye to our visitor and wished him a safe journey unimperilled by ghouls and phantoms.

Next morning, we were to visit Crom Castle, home of the Earls of Erne for the last 200 years. The canal crosses the demarcation line back into the North, and we came across a twelfth-century tower half a mile out on the lake. It was later used, apparently, by castle servants in days gone by for a one-night honeymoon if they got married. One night off work was all they were allowed.

We continued on to Enniskillen, and there it was that I met Frank, the genealogist who had been warned of our visit and had done some research. My Irish grandfather actor had styled himself rather grandly as C W Carleton-Crowe – it looked very good on the posters – and I, knowing my mother's maiden name was Crowe, assumed that he'd probably added the Carleton as an affectation.

No, said Frank, let's not bother about the Crowe part. Your grandfather's mother's name was Carleton, and she came from an Anglo-Irish family of some distinction, who

A fabulous round tower, on Devenish Island, once a monastic site.

settled here in Fermanagh, and, what's more, since 1613 had lived in a house called Rossfad, which stood just above on the shore of the Lough that we were due to pass through that very afternoon.

So, of course, we found the house, and met the present owner, a genial Scots doctor. It felt really uncanny, first of all to have been introduced to a whole family you never knew you had, in a place that you'd never been to, and then to find yourself sailing directly past the actual house in which that family had lived for four hundred years.

Pru, for a while, was rather respectful towards me when she learned of my elevated lineage; but it soon wore off.

We sometimes talk a bit about our early lives, the domestic conditions and our relationships with our parents; and here, for once, Pru's memory for detail is much sharper than my own. It seems she always maintained a close, loving relationship with her mother; and with her father, too, though she missed him very much during the war when he was stationed away from home. Her brother (also called Tim) died fairly recently. He also was in the army, and Pru says she had to search quite hard for any idea, socially or politically, that she could really share with him; but she loved him very much, and they would often talk together in detail about old times.

As a family they moved about quite a bit in the 1940s, although their spiritual home seems to have been a particular area of Surrey where her parents continued to settle in various different houses after Pru left home.

Coming out of Enniskillen, we had passed the school where Oscar Wilde and Samuel Beckett were both educated

(though not at the same time) and we got to thinking about our own schooldays. Pru went to Moira House in Eastbourne, was happy there, and, when the school was evacuated to the Lake District, she was accompanied by her mother, who had taken the job of under-matron. Although there was strictly a girls-only policy, for the sake of family unity Pru's brother Tim, then aged six, was allowed to join the kindergarten class.

So, all in all, a contented childhood, untroubled by doubts or feelings of inadequacy.

Mine was a bit different. We moved around a lot when I was very young, mostly because of my father's work as an itinerant actor, but also I suspect for pecuniary reasons. I don't remember being unhappy, though, or behaving in an insensitive way, until we settled in Bristol shortly after the beginning of the war.

I could tell that my mother was miserable. My father, now as a War Reserve Policeman, was on duty on alternate shifts of 6 a.m. to 2 p.m., 2 p.m. to 10 in the evening, and 10 p.m. to 6 the following morning; so a lot of the time when he wasn't working he was sleeping, and, in any case, not very good company. So it was hardly surprising that I would come across my mother crying quietly over the ironing board, and I believed that, in some way I didn't quite understand, I must be responsible.

My dad, with whom I used to enjoy games rearranging furniture to represent a river steamer (even in those days!) and taking the roles of captain and mate, no longer had the time to play with me, and I was left to my own devices. My parents loved me; I just don't think they *liked* me very much. Indeed I don't think I was very likeable. I used to fantasise about things I'd done, things that I'd made, or written, but that I hadn't at all.

I didn't have many friends, and when I got to Bristol Grammar School Preparatory Department, I avoided the attentions of my keenly flagellant headmaster by cycling off to Leigh Woods, or down to the docks at Hotwells to watch the shipping, or pretending to be some of the great figures of Bristol's past: Brunel, Cabot, W G Grace.

It never occurred to me to share my thoughts with my sister Patricia, who was five years younger than I; she went through school and passed on to university without occasioning, as far as I know, any domestic bother. I think our parents had enough trouble with me, and Patsy was rather left to make her own way.

Of course, I was sacked from Bristol Grammar, and this made family relations even more fraught. They hadn't improved, really, by the time I left my last school in 1950 and went on to the Polytechnic, Regent Street. Of course, I should have left home immediately then, but I didn't, and instead we went through all the familiar quarrels about staying out all night, about money, about unsuitable girlfriends.

They just *disapproved* of me. They hadn't liked Jacqueline, my first wife, at all, but when later I took the risk of bringing my friend Prunella Scales home to tea, my dad fell flat for her. I'm afraid we never achieved that feeling of affectionate interdependence that is so important when roles start to be reversed, and the sixty-year-old generation looks to the forty-year olds for support, but wounds became – almost entirely – healed, and they went on to become the most wonderful grandparents to Joe's family.

INTERLUDE
FROM THE LOG BOOK

Our domestic life at present naturally brings a few surprises, and sometimes I jot them down so that we can read them later, and enjoy them.

We have been sleeping in the same bed since, I think, 1969. It's an Edwardian brass job, and we love it, but its various bars and ornaments are held together by a pattern of screws, bolts and nuts, which occasionally come loose and bits drop off. Last night I was awakened by Pru twisting around, murmuring to herself. I asked what was the matter.

'It's – um – rattling around,' she told me. 'I can't get to sleep.'

Oh, dear! A bit's fallen off the bed. It does that. I listened, jiggled about a bit, but couldn't hear anything rattling. There's always a chance, though, that the bed might suddenly collapse without warning, so I got up, went down on my knees and touched all the little brass knobs. They seemed to be all in place. I felt on the floor. Nothing.

I crawled round the other side. Hannah, our cat, who sometimes chooses to sit on the bed, jumped down and started prodding me for attention. She's a compulsive stroke-seeker, so I have to spend a minute or two attending to her before continuing my search for brass nuts among the collected bits of history that seem to lurk beneath the bed.

Finally, I gave up. 'Bugger it!' I said. 'Can't find anything.' And I got back into bed.

'What were you looking for?' asked Pru sleepily.

I told her. 'Because you said something was rattling around.'

'My tooth,' she explained. 'The stopping's come loose. I'll ring the dentist in the morning.'

No brass bedstead on our narrowboat, I'm happy to say. Being so heavily involved with the series had meant that for a long time there'd not been much of a chance to get down to see her in her new moorings at Braunston in Northamptonshire and spend a few days aboard just by ourselves.

Getting to Braunston by road from London is boring, involving quite a lot of M1, but, once you're there in this historic boatyard, you can set off by water in any one of five directions: to Oxford, Birmingham, Coventry, Leicester, or down to London.

We got aboard, pushed off on the Grand Union, towards Warwick, and in the evening I decided to look through some of our old logbooks (there are now seven of them, dating from 1976 to the present day). I was reminded how the boat, being used by different members of the family at different times, and in different places, occasionally led to difficulties.

Friday 7th August 1989: TW, solus. Boiling hot day started well with early train to Banbury. Indian taxi driver picked me up at station to drive me to Nell Bridge, where Sam had left the boat. Driver talked all the time about how he knew Banbury, loved the canal etc etc, and duly dropped me at Nell Bridge.

I couldn't see the boat, had quite heavy suitcase, walked down towpath to where boat should apparently have been visible just above the lock, but no boat – and no lock either. Left suitcase in blackberry bush and walked on for about three quarters of a mile until at last I came to the lock. Which was not Nell Bridge Lock at all, it was King's Sutton Lock. Realised my know-all taxi driver had taken me to wrong bridge. Walked back, retrieved suitcase, found my phone, rang another taxi from the same firm (only one I knew) and eventually a confused lady arrived, with a <u>baby</u> in the back of the cab. She knew absolutely nothing about the area, but fortunately I had a map (she hadn't), and by making her go in the opposite direction every time she tried to make a turning, we eventually found Nell Bridge, but again no boat, and baby getting fractious.

It must be <u>somewhere</u>. Despairingly I set off in the other direction, and eventually saw it, tied to a tree stump. A note pushed under the door read, 'Boat has been moved as previous owner had left it opposite winding hole'. This is a serious boating transgression – a 'winding hole' being a point on the canal where it is possible to turn round – but I didn't think it should account for the loss of both mooring pins (now at the bottom of canal?).

Felt increasingly less paternally benevolent towards Previous Owner, who had collected some quite bad scratches to the hull, and left the fridge door closed on a full complement of mouldy eggs. And no log entry, either.

There were happier entries, like this one for Sunday, 16 July. I was involved in something called the British American Drama Academy. We held an annual summer residence in Balliol College, Oxford, and this year we were hosting a group of students from the Moscow Art Theatre School, with some of their directors and teachers. I decided to give a party for them all, on the boat. We got up very early and cast off from our Jericho boatyard to go through the Isis Lock and onto the Thames. The log reads,

Beautiful morning. Moored at Medley. Fetched and loaded Sasha Kalyagin, Irina Braun, Serbei Ostrovsky, Carolyn Sands, Caroline Keeley, Pru Keeley and two children, Tony Branch, Francesca Hunt, Brian Cox, Kate from the Office, someone else from the Office, Angus, Wendy, Phil (who?), Sally, Julia, Pat Mavroudis, Steffie, Derek, Juliet, Kate, Ben and some Russian children; and sailed them up to Duke's Meadows, where we gave them all picnic lunch with wine, plates, glasses and everything, and sat and watched the geese and the bathing cows and horses in their French Impressionist landscape.

Caroline's small nephew took all his clothes off and made a solemn collection of small dead fish culled from the Thames mud. Others paddled. One of the English-speaking Russian students told Steffie that he couldn't believe his eyes when he saw the river scene with cows and horses. It was just like pictures he had seen which always seemed to him to be of another planet!

Twenty-nine people on board! Of course, we were on the *river*: on the canal with that amount of human weight, we'd have hit the bottom and possibly sunk.

It was a really perfect day in every respect; and, in fact, if I were ever asked to pick out the happiest day of my life, I think this might just have been it.

A snap from, perhaps, the happiest day I can recall.

PART 5

SWEDISH
SAGAS

RECENTLY, PRU AND I were invited to a dinner – a black-tie affair – to be held at an address in Sussex for an organisation of which we are both patrons. To such events I don't drive the car, because I hope somebody there might offer me a drink; so I looked up a suitable train from Clapham Junction.

At 5.30, Pru took the curlers out and looked lovely in a long blue dress, I battled with my bow tie, and, looking like misdirected patrons for Glyndebourne, we got onto a train packed with home-going commuters.

We were the only people alighting at our countryside destination; in fact, when we got out, we were the only people to be seen anywhere. With growing unease we made our way down the road and through the silent, deserted garden of our rendezvous, and knocked at a door. It was half opened by an elderly, nervous-looking man in a woollen shirt, who eyed us suspiciously. Were we burglars?

We didn't *look* like burglars, but you couldn't be sure these days.

He shook his head. 'Tomorrow,' he said firmly, shutting the door.

It began to rain. We made our way back to the station and waited for the next train back to Clapham Junction. It was clearly all my fault: somehow, I'd put the wrong date in my diary, and, what's more, the next day, when the dinner actually *was*, we weren't actually free, so I was going to have to apologise.

We stood on the platform, wet and now quite hungry. Pru was extraordinarily nice about it. 'You just accept too many things to do,' she told me indulgently. That's true, but I'm afraid it's not just that. I do make mistakes about dates and times; I might easily say June when I mean July, or write 7.30 when I should have put 17.30.

So I ought to keep quiet about Pru's memory problem.

After that Sussex debacle, at least our next two or three weeks were accounted for. It had been decided that for our next *Great Canal Journeys* jaunt we should venture a little further afield – to Sweden. The idea was to cross the country from west to east, needing two programmes: one to cover a voyage across the western part of the country, from Gothenburg through the Trollhättan Canal and over Lake Vanern onto the Göta Canal; while the second part would take us further on, towards Stockholm and the Swedish archipelago.

It was an exciting prospect: I had worked in Norway and in Denmark, but never Sweden; Pru thinks she has probably appeared in Stockholm but isn't quite sure. We both knew a bit about Swedish folklore, and of course were well acquainted with the Swedish film-noir catalogue. Although it was July, we were warned not to anticipate summer weather: in Scandinavia, the general expectation is of meteorological gloom. Our new camera team of Chris Openshaw and Ben Crossley prepared themselves for a tough ride.

Sweden is a land mostly of water, with an island-strewn coast and a vast number of inland lakes. The Göta Canal, to which we were principally drawn, links two of those lakes and was created to enable vessels to travel from the Baltic in the east to Kattegat in the west. The Göta has fifty-eight locks to navigate, not all of them to be worked by us, but part

The oldest working passenger ship in the world, *Juno* made her maiden voyage in 1874.

of the route has apparently been christened 'The Divorce Ditch' after the navigational difficulties encountered by some married boaters. Hmm. We shall see.

We started off in style from Gothenburg aboard the *Juno*, a handsome, originally steam-powered vessel, the world's oldest registered passenger ship.

She was launched in 1874, rebuilt in 1904 and converted for diesel in 1956. The full journey from Gothenburg to Stockholm takes four days and three nights, but we were going only as far as Sjötorp, with one overnight stay in a double cabin (double in the sense of upper and lower bunks, not easy for the one on top – me – to get down in the middle of the night, in the dark, for whatever reason).

We had breakfast, and transferred to our hired 36-foot cabin cruiser with the nice ability to steer from either inside or up on the flybridge (the open area on the top of the ship), and set out along the Göta.

A toast to the Göta.

At Forsvik, a wonderful surprise was awaiting me: a Victorian paddle steamer! The *Eric Nordevall* is a replica of the ship that sank in Lake Vattern in 1856. The wreck was discovered in 1980, and it was decided to build a perfect reproduction, powered by a single vertical cylinder and fired up, as in the original, by wood rather than coal. We came aboard for an hour's sail, which was bliss for me; and Pru had a nice time too with the vessel's young captain, very dishy in his nineteenth-century tail coat.

There is a floating sauna at Karlsborg, and I persuaded Pru to try it. She says she had never been in one. I was reminded of a health club I once belonged to (what was I thinking of?) whose patrons included a quite famous comedian with a very bad wig. He knew it was terrible, and would go into the sauna only if he could sit on the top step so that nobody could see the top of his head. I'm glad I don't have that problem.

We enjoyed the sauna, but passed on the cold dip.

Records show that a great east-to-west canal had been discussed in Sweden since the sixteenth century, when the nation was under Danish rule. But it was not until 1810 that King Karl XIII of Sweden commissioned Admiral Count Baltzar von Platen to build the Göta Canal, the greatest civil-engineering product undertaken in Sweden at that time, and necessitating the assistance, as overseer, of our own Thomas Telford, who brought with him British engineers, craftsmen and equipment.

Such a colossal project naturally took time to complete, and it finally opened in 1832, by which time, of course, railways were being planned along much of the course.

Navigating the Divorce Ditch's infamous Berg Lock with marriage intact, I am glad to say.

When they finally arrived in 1855, the effect on the canal was disastrous: not only could trains carry passengers and freight more speedily, but there was no need for them to shut down in winter (the canal was impassable for five months of the year).

We approach the end of the canal: the famous Berg Lock flight. This is it, the final slope of the Divorce Ditch. How will we survive it, physically, mentally, conjugally?

Actually, we did – calmly and methodically, just passing the ropes through the correct rings, being civil to each other, and in fact enjoying the adventure.

So we passed on to Part Two of our Odyssey. This was to take us across Lake Mälaren to Drottningholm and Stockholm, and the islands beyond. We went through commercial Södertälje canal onto the lake, created, according to Norse mythology, by the goddess Gefjon when the Swedish king

Gylfi promised her as much land as four oxen could plough in a day and a night. She used special oxen from the Land of the Giants, who uprooted the land and dragged it into the sea somewhere, leaving a space that became Lake Mälaren.

Swedish mythology is full of such tales, and these stretches of water, silent but for the occasional cry of terns, mallards, gadwall, cormorants and sandpipers, impart an element of fantasy peculiar to these surroundings.

The feeling of sailing across a lake is in any case a different experience from following the course of a river or canal, where you are looking at the bank and observing nature close-to. On a big lake such detail is too far away, and you are left to your own thoughts.

How good are we at reading each other's thoughts under such circumstances? As actors, we have to learn to make audiences aware of our *own* thoughts when necessary, without the assistance of language; but that's not the same as being able to read someone else's mind, when they're looking into the middle distance. As I sat on the bridge of our boat, steering across Lake Mälaren, Pru was just a yard or so away; and it occurred to me that I didn't have the slightest idea what was going on in her mind. Was it the pleasure of the moment, to be in this sublime habitat? Was it reminding her of something in the past? Or was she thinking of the future?

We don't talk about the future, for perhaps obvious reasons. For someone with any form of continuous dementia, the prospect is not cheerful. Pru seems to cope wonderfully with things day by day, as I try to; but surely she must sometimes give a thought to whether she is

eventually going to be allowed to end her days in the house that has been so dear to us for the last forty-five years.

Pru and I don't concern ourselves with death, although of course we have sometimes talked about the possibility of an afterlife. We are both Church of England members, and Pru says she thoroughly believes in the hereafter. I have problems with that, and suspect the biblical promise may be intended to have a subtler, more indirect meaning. What comforts me about dying is the thought of my own insignificance: the birds in the garden will go on singing, the grandfather clock will continue to tick and the 219 bus will stop outside the house as usual.

While we were riding in a traditional Viking ship, Pru read me a verse from the *Sturlunga* saga, an Icelandic Viking poem:

> Old man, keep your upper lip firm
> Though your head be bowed by the storm;
> You have had girls' love in the past.
> Death comes to us all at last.

Well, thanks.

We came to the island of Birka, a Viking settlement in the eighth century, and learned that, contrary to popular belief, 'Viking' meant that your occupation was as a seafaring merchant and not just a bloodthirsty warrior. There would have been about a thousand people living in Birka, a well-planned town with wooden housing on stone foundations.

Archaeologists, having excavated 5 per cent of the island, stopped: they had found out so much already that further

Opposite: Sailing in a Viking ship across a remarkable lake. I realise that sometimes, although we may travel beside each other, there is a ford between our thoughts that cannot always be brooked.

215

Drottningholm palace, outside Stockholm, was of particular interest to us as it is the home of the oldest preserved theatre in the world. © Getty

digs would be unnecessary. Among the findings at nearby Rastaholm in 1954 were a small Buddha statuette from northern India and a christening scoop from Egypt. They got around.

On this journey there were so many places to stop, and explore, and film, but we just didn't have time. And remember, of course, that not everything that we do record makes it to the final edit. I believe, though, that we somehow managed to capture something of the atmosphere of this remarkable country. When we met the Swedish actor Krister Henriksson, who plays Wallander in the TV series, he was able to explain something of how the Swedes react to our own conception of them.

'We behave in the way you want us to,' he says. 'Melancholy, lonely, depressed. But loneliness is attractive.'

Drottningholm, though it boasts a handsome royal palace built in 1580 and rebuilt in 1662, is chiefly of interest to us because of the historic theatre standing behind it. Built in 1766, the same year as the Theatre Royal, Bristol, it carries the distinction of being the oldest theatre in the world in its original state (Bristol has been remodelled in detail from time to time). All the original technical machinery remains in working condition beneath the stage.

Going on to Stockholm, and after wandering through the beguiling city, we were at sea again, making for the little island of Sandhamn, crammed with visitors in the summer but with a resident population of fewer than a hundred souls. We were entertained at a crayfish party by a successful novelist who has nevertheless managed to write eight murder mysteries about the inhabitants.

Indulging in Crayfish in Sandhamn – where the fictional character Mikael Blomqvist kept a cabin in Stieg Larsson's global bestselling *Millennium* crime trilogy.

STRATFORD-
UPON-AVON

PRU GAVE HER FIRST performance at Stratford-upon-Avon in 1956, when it was simply the Shakespeare Memorial Theatre, offering its annual season, and before it became the home of Peter Hall's Royal Shakespeare Company. My own first visit to Stratford, though, was on the tour of *Simple Spymen* – rather different.

I didn't come up there again until the RSC season of 1965. We rented a cottage in Tredington, a few miles out of Stratford, and it was here, I suppose, that Sam was conceived. Pru was doing *The Winter's Tale* at Birmingham Rep, an easy journey from Tredington by car; we brought our cat, and had a very happy nine months.

Some of the rest of the company, I think, were not so fortunate. The Stratford seasonal pattern familiarly unfolds in this way:

In March, you travel up to Warwickshire, explore the town, meet your colleagues and look out at the Avon sparkling

in the spring sunlight. You're in perhaps five out of the six plays, and the one in which you've got the best part should logically, you think, be the one most applauded by the critics. The company are fielding a cricket team, so you put your name down. You've rented a nice little cottage, which is just across the street from a very attractive girl on the stage management. Yes, it looks like a good summer.

The season draws on, and now it is October, and things are coming to an end. There is fog on Clopton Bridge. Your favoured play was universally disliked by the critics; you were not mentioned. The attractive girl went off with the electrician, the cottage roof leaks and the cricket team lost every match.

It wasn't quite like that for me, and I had a fairly good season; but I did want to know what the company planned to do with me next year. Not much, it seemed. So I left (probably a mistake), and didn't return to Stratford until 2007, in *Coriolanus*. This was the last play to be shown in the Royal Shakespeare Theatre. At the end of the run it was to be pulled down and the name given to the much-vaunted three-sided auditorium that would take its place.

The old theatre, built by Elizabeth Scott in 1932, had never been hugely popular with audiences or actors, and on the closing night the public speeches were of buoyant expectation of the future, rather than grateful valediction to the past.

I was a bit sad, though. When a theatre is demolished, something dies with it: the ghosts who have stood on that stage over the years, the legends that have been born there. There are quite a few of those; Shakespeare nourishes them.

Opposite: Pictured as Falstaff in Shakespeare's highly popular history, *Henry IV*.

But, for the present, here we were, on our boat, drifting along this picturesque canal, trying to recapture our initial feelings of excitement at being allowed to speak the words of the Bard here in his hallowed birthplace.

I have, I think, performed now in twenty of what are now generally accepted as the thirty-seven plays he wrote or had a major hand in, and I've done some of them two or three times. I can't think of anything more enjoyable professionally than to go on performing his work for the rest of my life.

How did I come to love him? I think the credit belongs to my English master, 'Sammy' Cowtan, at John Lyon School, where I went when the family moved to London.

'Look, you're never going to understand this stuff unless you get up and *do* it,' said Sammy. 'Unless you know what it feels like to say those words, and to have them said *to* you. Right, clear all the desks to the wall; you're Brutus, you're Cassius, you're Casca. Come on, louder!'

If we were reading love scenes, Sammy would insist on playing the female parts himself. Even at the time, I remember wondering if he had any psychobiological reason for doing this; but no. It was simply to avoid the embarrassed giggles that would go on if the scene were being played between two boys (it was a single-sex school).

He was very good, too. Bald, corpulent, bespectacled and in his sixties, Sammy was nevertheless an attractive and entertaining Viola, a powerful Cleopatra and a really scary Lady Macbeth. I don't think his Juliet really showed him at the top of his form, but I believe it had its adherents in the class.

But back on the boat. We had joined the Stratford-on-Avon Canal at Kingswood Junction, continuing south through Wootton Wawen and over the iron Edstone Aqueduct. We picked up my daughter Juliet, granddaughter Kate and, yes, great-grandson Kaya on the way, and picnicked on the bank in Shakespeare's Forest of Arden. Arriving at Stratford's Bancroft Basin and passing through the lock, the River Avon stretches away to the right, past the theatre and Holy Trinity Church, and on towards Evesham and Tewkesbury. To the left, the Upper Avon takes you beyond Alveston and Charlecote Park to where it soon becomes too shallow to navigate.

Picnicking with grandchildren – and great-grandchildren – on the banks of the Avon, in the heart of Shakespeare's Arden country.

There are plans, which I enthusiastically support, to extend this navigation up to Warwick, thus linking the two colossal tourist centres directly by water. At present if you want to do the trip, you are faced with a thirteen-mile, thirty-eight-lock journey from Stratford up to the Grand Union at Kingswood Junction, with another seven miles and twenty-one more locks before you get into Warwick. The extension is one of the things I hope I may see happen before I die. (Allowing women members into the Garrick Club is another.)

While Sam was up in Stratford, playing Richard II one year and Hamlet the next, we lent him our boat so that he

The Dirty Duck is a favourite battle cruiser to legion Shakespearean actors, thanks to its proximity to 'the office'.
© Rex Features

could live on it for the season. He managed to get the RSC's private mooring on the river, outside the Black Swan, known by actors for generations as the Dirty Duck. Not a bad way to spend your summer, I thought: playing Hamlet till 11 o'clock, into the Duck for a lock-in till 12.30, and then, 25 yards across the grass and you're home, if you don't fall in.

I rang him a few weeks into the season to ask him if everything was all right, and he told me the water pump was not working. I told him where to ring to get a replacement; you can't live without water, I told him.

'I know, it's very inconvenient. I can't make ice for my cocktails.' (Social life has come on a bit since I was at Stratford.)

When Sam had a bit of time off while a new show was being opened, we borrowed the boat back for a fortnight to go down the river, past the picturesque Avonside villages of Welford and Bidford to Evesham and then Tewkesbury, where it joins the River Severn. We went up the Severn to Worcester, where we spent a little time, then turned onto the Worcester and Birmingham Canal to get back onto the Stratford at Kingswood Junction. A lovely, not-too-pressurised 'ring' voyage, much to be recommended.

VENICE

GREAT CANAL JOURNEYS had provided us with a busy spring season, and now it was nearly time for our next annual commitment: the BBC Proms season. We love the Proms, not just as a chance to hear some wonderful music at the hands of the best performers in the world, but as a celebratory come-together of every age, race, creed and social background. Whether you like Bach, Zemlinsky, Jools Holland, Charlie Mingus or Pacific Island nose-flute music, there will be something to delight you. The Royal Albert Hall is a fantastic building to be in. And it's *cheap*.

I don't really know how Pru and I each came to love classical music. It didn't come from our parents. My father bought Parlophone 78-r.p.m. discs of Victor Silvester and his Dance Orchestra; Pru's family stocked up on songs from the shows of the thirties, featuring people such as Jack Hulbert and Cicely Courtneidge. We certainly didn't have a piano, either of us.

Both of us were born just too early to get caught up in the pop revolution. So I'm afraid we never really found out what pop music was about. As students we went along with what our slightly older colleagues were keen on, and for me that was Sibelius and Humphrey Lyttleton.

Of course, I appreciated the Beatles and the Stones and one or two others, but they were special; they stood out because of their talent and individuality from the unasked-for noise that in those days was going on all around me, some of it very good but which I didn't accept as being 'music'. It's hard for the two generations below me to understand how completely ignorant I am about pop music (or Music, as it's now called).

I went in the other direction quite keenly. We had the radio on a lot in my early days, and heard a lot of very good music. I went to concerts, and to opera when I could afford it, and I learned to read music. We could never afford a piano, so I never learned to play, but I joined local choirs and enjoyed singing oratorio. Nowadays, I seldom get the chance to sing professionally, but I think that is because singing roles often require one to dance a bit, and for me that's out of the question. I am unquestionably the most uncoordinated hoofer in the business. No, really.

Pru's story was slightly different: she actually studied music as a subject at school, did a lot of singing during her drama training (and dancing, of course) and performed quite a bit in intimate revue.

Her love of music – particularly of Handel opera – was one of the things that first drew me to her in our early days; and even now, in her present state, where she can enjoy

seeing a play but can't tell you much about it afterwards, Pru will give herself completely to a piece of music, and follow its mood and structure.

We had three more programmes scheduled for August into September. By this time quite a lot of people were stopping us in the street, telling us how much they were enjoying the series (how much their *mothers* were enjoying the series was a frequent comment) and suggesting places for us to go next.

'You haven't been to *Venice*, they suggested. 'There are lots of canals there.'

Well, yes, we know, and so many other things, too, but

We adored Venice. It's a place everyone should see once in their lifetimes – its beauty is unparalleled.

let's leave our visit to 'The Pearl of the Adriatic: the Bride of the Sea: the City of Romance' as late in the year as possible, when the tourist season is tailing off a bit. . .

For Venice now, the human menace of mass tourism outweighs the natural peril the city has always had to fear of being washed away into the sea. The historic old city's residents (55,000) feel their quality of life is fast being plagued by massive overcrowding, misbehaviour, vandalism and crime. Many can no longer afford to live in their homes because of inflated real-estate prices being elevated to the apparent spending power of the thirty million annual visitors.

The big cruise ships are the worst offenders. Docking in the Venetian lagoon, they disgorge an enormous number of passengers who are fed their meals aboard and so don't even contribute revenue to local restaurants. The city authorities are apparently unapologetic about welcoming these huge vessels, as 'Venice keeps the entire Adriatic cruise industry afloat, and provides 5,000 jobs.' Yes, but things are now so bad that its status as a UNESCO World Heritage Site is now under serious threat.

The *GCJ* team arrived, and tried very hard not to behave like dreaded tourists, although our original plan was to arrive on the Orient Express to Verona, and thence to Padua to approach Venice via the Brenta Canal. I'd been on the Orient Express once, on my own, and on a one-way trip from Venice back to London. This was some time ago, and there was a difficulty, which I'm sure they've now solved, about keeping the batteries charged while the train was held in a siding.

As you were on tracks actually owned by Italian, French and Swiss Railway companies, it was reasonable that you had to let their crack trains pass every so often, and, although your schedule easily permitted this, what happened was the fans would stop revolving, your lights would dim, your ice would melt and catering facilities impeded. Otherwise, fine. It was an experience.

In the present situation, it was decided not to come by rail, and we flew into Venice to stay our first night in Mira, on the mainland; and next morning met our new DOP (Director of Photography) Steve Robinson, with his son Jake on B Camera, at the ancient little town of Dolo, where our boat was waiting to take us along the Brenta Canal.

We might have tried not to look and behave like the tourists that flood the city every year. But I expect we failed.

Above: The Brenta river, which takes one from Padua towards the Ventian lagoon, is the most stunning scenic entryway to this most magical of cities. © *Getty*

Right: Built in 1569, the Ponte degli Alpini bridge, in the town of Bassano, overlooks the Brenta river. During World War II, the city was invaded by German troops and the bridge destroyed. Thankfully, it was rebuilt in all its former glory. © *Getty*

This important link from Padua to the Venetian lagoon first came into being in 1209, but its original purpose was to control the Brenta River, which regularly flooded fields and crops and carried silt from the lagoon into the existing water channels.

The six locks on the canal are thought to have been specifically designed by Leonardo da Vinci, but they lay upstream of our journey, so we didn't see them. What we did see were an imposing number of very grand houses built along the canal as a result of the repeal of an early law forbidding Venetians to build on the mainland.

One of these is the Villa Foscari – also known as La Malcontenta – built in 1555 by Andrea Palladio for the ancient noble Foscari family, who have owned it since that time. We were privileged to meet Antonio Foscari, who rescued the place from neglect in 1973 and restored it to its original condition. Why Malcontenta? Well, there is a story that a Foscari wife was imprisoned here for a long while because of a social indiscretion, and was understandably malcontent.

Coming out of the canal into the wide lagoon, there is a marked channel directing boats across to the city of Venice itself. Pru and I had been to Venice a couple of times before, but never approached it from the mainland, as we were now doing. We looked out into the distance and watched the gradual appearance through the mist of this magical model island sitting in the sea. It was an unforgettable experience.

Most cities in the world are in a good, accessible location. Venice was settled on deliberately because of its *bad* location: a haven and hiding place from barbarians who lacked ships and a knowledge of the sea. The city, founded on hammered wooden poles driven into the seabed and now petrified, grew in the fifth century to a comfortable independence allowing their isolation from Italian political life and the creation of what was to become the great mercantile empire of the Venetian Republic, and the most dazzlingly beautiful city in the world.

We at last bullied our way into a mooring-place (how well we know that problem) and disembarked onto the Piazza San Marco, the social, religious (and of course touristic) centre of Venice. At the Caffè Florian, patronised at different times by Byron, Goldoni, Goethe, Dickens and Proust, we were told of the strong aphrodisiac properties of hot chocolate, customarily ordered for his companion by Giovanni Casanova.

Pru, Juliet and I had been invited to stay at the home of two cousins who let out part of their *palazzo* on the Grand

The manager of the historic Caffe Florian throws open the door, which looks out onto the stunning St Mark's Square.
© Getty

Canal to visitors, while retaining one floor for themselves. This has now become common practice in Venice.

Very comfortable – a great deal more so than the *topetta*, the small outboard boat in which I had to steer us through the canals. From an uncomfortable perch I had to put one hand behind my back both to steer and regulate speed. Resident Venetians (there are still a few of them about) manage this with skill and grace, but not I, I'm afraid. Our wonderful local guide, Marco, was kind about my performance, but he didn't see me trying to get the boat through a narrow canal to the stage door of the Teatro la Fenice (one of the very few theatres in the world where the 'get-in' of the scenery is by water).

I had performed *King Lear* at la Fenice in the Venice Bienalle of 1971. Since then, the exquisite theatre was destroyed in an arson attack, reconstructed at a cost of €90 million and rose triumphantly like the phoenix (*fenice*) from the ashes in March 2001. It was enormously exciting to have played here.

A different sort of theatrical experience awaited us at the Nicolao Atelier, where the actor Alessandro Bressanello dressed us in traditional costumes of the Commedia dell'arte and improvised some scenes with us. It's a very actor-based kind of theatre; there is no place for writers or directors. But Goldoni, while using familiar Commedia situations, developed them in a more formal style, with the playwright in control.

It's interesting, though, that Commedia troupes were actually the first in Europe to employ women to play female parts. This device proved so popular with audiences that sometimes scenes were included that encouraged girls to appear naked (in a shipwreck, for instance, or a fire). English critics were not impressed by this: Ben Jonson wrote of one female Commedia performer as a 'tumbling whore'.

We visited the Arsenale, the great shipyard at which, in the eleventh century, shipwrights, using separate workshops for carpenters, sailmakers, pitch-boilers, ropemakers and so on, could construct up to three large ships in a single day. The present, enlarged, complex was built in 1320.

Dante, in *The Divine Comedy* pays tribute to the tireless workforce:

> One hammers at the prow, one at the stern;
> This one makes oars, and that one cordage twists,
> Another mends the mainsail and the mizzen.
> Thus, not by fire, but by the art divine,
> Was boiling down below a denser pitch
> Which upon every side the bank belimed.

These Venetian galleys were light and agile, not built to take loads of freight, but to accommodate a crew of captain and experienced seamen, crossbowmen, a mapmaker, a scribe, a doctor, the rowers, and even a few civilian merchants. They made Venice the mercantile capital of the world.

We went over to the island of Burano, where we met two attractive and muscular young women who were the local champion rowers – rowing being an important sport and accomplishment in Venice, in a very particular form. In their boat of individual design, they stand with an oar each, one in the centre of the boat, one at the stern, and speed

A former naval base dating back almost 1,000 years, the Arsenale was the centre of Venice's nautical might. When, in the sixteenth century, Turks attacked the Cyprus, the Venetian Arsenale was able to provide a small army of boats in a matter of weeks. © *Getty*

along at an incredible rate. The girls kindly asked me if I'd like to have a go, but still recovering from trying to keep us alive in our *topetta* in the open waters of the lagoon, dodging between all sorts of craft out for the day, and being hooted at by oncoming *vaporettos* (the water buses that ferry people around the city), I thought I'd leave it to the experts.

They took us across to Torcello, where the original Venetian settlers made their homes, and it was here in the shadow of the ancient Cathedral of Santa Maria Assunta that we said our farewell to Venice.

However, while pretending as a film unit to be a cut above the vulgar tourists, of course we finally succumbed to their timeless image: a ride in a gondola. Late at night, when the Grand Canal was deserted, Pru and I sailed in the moonlight. Just the splash of our gondolier's oar, and his song echoing from the walls of the silent palaces – it doesn't get much better than that.

JOURNEY'S
END

IN TERMS OF THE SHOW, that brought us to the end of 2015. Now we had to think carefully about the future. Audience figures had continued to climb: Channel 4 were very keen for us to continue, certainly for another year. A visit to the Netherlands had already been researched, two more Scottish canals had been talked about, and I was very keen to do a programme on my home town of Bristol, with its very unusual pattern of water management.

Michael was up for it, with the whole Spun Gold team behind him. It really depended on Pru.

Looking back over the very first few episodes, I could detect that she had now lost something of the ability to spring around like a mountain goat, but come on! After all, she was now eighty-three. The important thing was that she loved the programme and was still interested and involved in what we were doing.

Everything took a little longer now – of course, Pru's

memory was failing – but we all had to recognise that if we were prepared to be patient, everything would turn out right in the end. The team were wonderfully generous and supportive, and I don't mean just about Pru: my own speed and ability was sometimes showing clear signs of wear and tear.

Anyway, we carried on, for the next couple of years. We went on to make episodes travelling the Caledonian Canal through Loch Ness and being marooned on the Isle of Islay in bad weather. We went to Portugal, to Alsace, to the Netherlands, and spent two wonderful weeks in India: to Kerala in the south and Assam in the north-east. Occasionally we had to cheat our title a little: the Brahmaputra River can scarcely be described as a canal, but it is a very noble and vital waterway that has carried boats full of merchandise up and down its course for centuries, so it does all that a canal should.

Now we have shot the last four episodes: the Canal de la Marne au Rhin in Alsace, the River Douro in Portugal, the Norfolk Broads (surprisingly), and the Brecon and Monmouth Canal in Wales. I said 'surprisingly' about the Broads, but in point of fact every single waterway we have covered in the series yielded surprises of some sort or other.

One of the surprises, of course, has been the continuing success of the programme itself. When we started out with our first four offerings in August 2013, we imagined a few old ladies turning to More 4 on a wet Tuesday afternoon; but the audience figures have steadily grown, and with them our confidence in exploring new avenues of adventure.

As this book goes to press, we have now shot the last scene of the last episode, so that is the end of *Great Canal Journeys.* It is really sad to be saying goodbye to such a wonderful group of people: Mike, Trina, Chris, Dave, Julian, Steve Robinson and Barney Carmichael: a great team, and we've been so happy. All things come to an end though, and as they say, 'Quit while you're winning.'

Off camera, we will now get back to our own beloved boat, in Braunston Marina (you probably know the traditional Australian ballad: 'Braunston Marina, Braunston Marina, Who'll come to Braunston Marina with me?') and take off for a few days, just pottering along and tying up somewhere to read a book and open a bottle of wine.

A love of boats and sailing truly has been a thread through our marriage, since the very beginning.

I wonder how many of our viewers have been turned on to the idea of hiring a boat and trying a week or two on a canal that takes their fancy? We certainly get letters from some people who've got hooked, and that's great to hear. I'm sure there have also been disastrous maiden voyages, but people are perhaps too nice to tell us about them.

A bit of first-time training, about steering, about canal etiquette, about the use and function of locks, is essential. Sometimes what advice you get from the boat-hire people is fairly minimal; that's because on a Saturday morning they may have fifteen of their boats returned to them, which they have to clean, tidy up, empty the loo and fill up with diesel and water before they're ready to go out again. That doesn't leave a lot of time for instruction to the new hirer:

'Look, squire, it's dead easy really: this is the tiller, push it left when you want to go right, right when you want to go left. This is forward, that's reverse, that's neutral. Pass other boats on the right. Good luck, have a good week.'

I've written about the dangers of positioning yourself badly in a lock when you're going down, and getting caught on the cill; but otherwise it's just common sense. Take it easy: don't let the water in too fast.

In fact, 'take it easy' is good advice for the whole trip.

I'm sorry we couldn't get all our journeys into this book. I have loved them all, and I really believe Pru has, though it's difficult now for her to talk about any of it in detail.

Except from occasionally feeling irritated that she's not been able to recall some person or occasion to mind, Pru remains blessedly free of depression and distress about her condition. She is grateful for all the good things she's had, and still has, in life: the love of friends, of her family and me, a house and garden that she treasures, and a rich life to look back on.

This morning, she got out of bed, drew back the curtains and looked out over the sunlit Wandsworth Common.

'It's going to be a lovely day,' she told me.